FOOD SAFETY REGULATORY COMPLIANCE

Catalyst for a Lean and Sustainable Food Supply Chain

Series on Resource Management

Food Safety Regulatory Compliance: Catalyst for a Lean and Sustainable Food Supply Chain
by Preston W. Blevins
ISBN: 978-1-4398-4956-9

Driving Strategy to Execution Using Lean Six Sigma: A Framework for Creating High Performance Organizations
by Gerhard Plenert and Tom Cluley
ISBN: 978-1-4398-6713-6

Building Network Capabilities in Turbulent Competitive Environments: Practices of Global Firms from Korea and Japan
by Young Won Park and Paul Hong
ISBN: 978-1-4398-5068-8

Integral Logistics Management: Operations and Supply Chain Management Within and Across Companies, Fourth Edition
by Paul Schönsleben
ISBN: 978-1-4398-7823-1

Lean Management Principles for Information Technology
by Gerhard J. Plenert
ISBN: 978-1-4200-7860-2

Supply Chain Project Management: A Structured Collaborative and Measurable Approach, Second Edition
by James B. Ayers
ISBN: 978-1-4200-8392-7

Modeling and Benchmarking Supply Chain Leadership: Setting the Conditions for Excellence
by Joseph L. Walden
ISBN: 978-1-4200-8397-2

New Methods of Competing in the Global Marketplace: Critical Success Factors from Service and Manufacturing
by William R. Crandall and Richard E. Crandall
ISBN: 978-1-4200-5126-1

Supply Chain Risk Management: Minimizing Disruptions in Global Sourcing
by Robert Handfield and Kevin P. McCormack
ISBN: 978-0-8493-6642-0

Rightsizing Inventory
by Joseph L. Aiello
ISBN: 978-0-8493-8515-5

Handbook of Supply Chain Management, Second Edition
by James B. Ayers
ISBN: 978-0-8493-3160-2

FOOD SAFETY REGULATORY COMPLIANCE

Catalyst for a Lean and Sustainable Food Supply Chain

Preston W. Blevins

CRC Press
Taylor & Francis Group
Boca Raton London New York

CRC Press is an imprint of the
Taylor & Francis Group, an **informa** business
A PRODUCTIVITY PRESS BOOK

CRC Press
Taylor & Francis Group
6000 Broken Sound Parkway NW, Suite 300
Boca Raton, FL 33487-2742

Printed in the United States of America on acid-free paper
Version Date: 20121011

International Standard Book Number: 978-1-4398-4956-9 (Hardback)

Visit the Taylor & Francis Web site at
http://www.taylorandfrancis.com

and the CRC Press Web site at
http://www.crcpress.com

Contents

Foreword

This foreword is different. It consists of five short forewords from individuals who have different areas of expertise and are respected "thought leaders." The contributors to this foreword include an industry analyst and journalist, an authority on best practices in operations and supply-chain management, a chief information officer (CIO), a respected guru on the Lean Management System, and an expert on sustainability best practices.

* * *

The food and beverage industry comprises producers and processors from the farm level to the multi-billion dollar brand owner corporation who want to sell safe and nutritious food at a competitive price in the new globally connected ultra-competitive economy.

With the recent enactment of the Food Safety Modernization Act and the demands by large retail chains that their food and beverage suppliers become GFSI certified, all processors face rigorous food safety challenges while trying to understand their business—from bottom to top and top to bottom. The challenge is to efficiently improve food safety while continuously improving operational productivity and supply-chain reliability, and moving toward a sustainable green business model.

There are three constraints to achieving the desired result:

Information systems that do not support both the "bottom to top and top to bottom" understanding of operations and supply-chain status
Wide use of Enterprise Resource Planning (ERP) systems that were not designed to support the special needs of the food processor
Inadequate dissemination of best operating practices across supply-chains

Today's processors need to have actionable business analytics, understand how their supply chains work, comply with all government regulations that apply (sometimes international), create innovative food and beverage products, reduce the consumption of energy and water, and practice food safety at all times. Over the years, various software systems have sought to help processors understand the complexities of their businesses—though in many cases these systems were adapted to the food industry from general-purpose software systems, and did not integrate well with other suppliers' systems.

While the software industry has helped many large food and beverage companies achieve their business, quality, and efficiency goals, by integrating (behind the scene) disparate systems, it has no doubt confused many small- and medium-sized operators with promises that did not pan out—largely because the software supplier did not fully understand the special needs of a regulated industry.

This "hands-on" book pulls together in one place many of the loose ends that processors may have neglected—only because they have been trying to get safe food and beverage products out the door. The book walks the reader through food safety issues, product traceability and supply chains, labor issues, maintenance, automation, training, systems integration, and much more. Within each of these topics, food safety and regulatory issues are interwoven, providing the reader with the ability to see the big picture, and to understand that the sum of the parts is often greater than the whole.

Many books, for example, deal with just quality or food safety issues, and others with Lean manufacturing, and still others with supply chains, and so forth. This book brings together the information a reader needs to see the big, integrated picture, enabling him/her to "walk the walk and talk the talk." Reading this book will help arm the processor with actionable information, such that it cannot only make the right demands of its ingredient suppliers, but also its automation and software suppliers. At the same time, the book will help the reader understand how to apply systems that can help with food safety, quality, and operating efficiencies while meeting the demands of its customers—the retail chains and big-box stores or maybe even the direct retail market. Quizzes at the end of each chapter help the reader understand where brush-up work is needed.

Wayne Labs
Senior Editor, *Food Engineering Magazine*

* * *

Preston has provided a very compelling case on how to ensure the availability and quality of our food supply goes beyond its current state. Through the use of proven methodologies practitioners, students and the industry as a whole can use this book as an invaluable resource to effect the necessary changes. A clear example is the use of Lean methodology; Preston takes the reader through the development of a clear understanding of the current state and statistical analysis of the key performance indicators, more importantly the value of clear and accurate data. Once developed, the process review follows the products through the various departments, explaining how workflow diagrams can be used to identify areas for improvement both short and long term.

Preston has provided thoughtful and expert advice using well-established process improvement techniques to guide the reader on a most critical journey. This is must-read for all stakeholders in the food industry.

Abe Eshkenazi

Chief Executive Officer, APICS

* * *

Preston set out to make a complex topic, the food supply chain and the influences that shape it now and will shape it in the future, understandable and approachable to the world of the mid to small food manufacturer. In his thoughtfully prepared book, however, he has accomplished that and much more. His easy-to-read explanations of the many terms, formative concepts, and particularities known only to those deeply acquainted with the food supply chain framework, make his book one of universal value to *anyone* concerned with the current state of our food supply processes and what we need to collectively do to favorably influence them and *our* future.

But, he didn't stop there.

He forthrightly points out the "elephant in the room," that we all know is there: sustainability. From reduced carbon footprint to blue water, the need for it, especially in the food supply chain, is no longer an "if we can" option. Preston hammers home the point that sustainability in our food supply chain will imminently be a requirement for all food manufacturers concerned with their long term viability and he gives numerous global examples of food and nutraceutical firms (i.e., Nestlé and Bayer) that have 20-year sustainability programs in place and are commercial successes (underlining the obvious fact that these two things are not unrelated).

In short, I couldn't recommend this book more highly to anyone trying to make sense out of and understand the happenings in the food supply chain. But I don't mean this only from a yesterday and today perspective. Preston sets forth how we can introduce ever more beneficial change to realize tomorrows that, at some level, we all want to see.

Al Kuebler

Chief Executive Officer, Technical Impact

Lean Thinking is an approach to process management and process improvement that has been getting widespread traction over the last decade. With its roots in manufacturing, Lean is now widespread in healthcare, the military, construction, nonprofits and government. One thing that practitioners are learning is that regardless of industry, Lean is a relevant approach to improving performance. And within any organization, most all processes will benefit if looked at from a Lean perspective.

In this chapter [7] Preston presents to the reader a broad set of thoughts, behaviors and tools that help those within the food industry understand what "Lean" means, as well as how the Lean journey can support improvement needs in the areas of service, quality, cost, safety and morale. We start from a strategic perspective to help explain the context of Lean; Lean is explained as a philosophy that guides an organization through customer-defined value, voice of the customer, developing thinking problem solvers, and value stream management. With this foundation we then delve into the core Lean tools such as visual management, kaizen and kaizen events, cells, and standard work.

In his review of each concept, Preston provides an overview followed by management considerations. This all serves as a great foundation to explore the power and relevance of Lean within the food industry.

Mike Osterling, author
The Kaizen Event Planner: Achieving Rapid Improvement in Office, Service, and Technical Environments

* * *

Provided the ever-evolving challenge of defining Sustainability for a dynamic field such as Food Supply Chains, Preston strategically frames his discussion by clearly establishing appropriate taxonomy based on the Kyoto Greenhouse Gas Protocol and acknowledging the conceptual convergence of overarching paradigms such as the triple bottom line, the three dimensions of sustainability, and their balanced scorecard metric equivalents.

This chapter [8] provides an orientation to some critical aspects surrounding sustainability management by describing the standards on Greenhouse Gas (GHG) and water management that the Global 500 and their supporting supply chain partners have adopted. Particularly, it describes how the SCOR Framework, the GHG Protocol, and the Water Footprint Network timely address the three dimensions of sustainability.

Readers will understand how the Lean Management System and the Water Footprint Network methodology can be applied to design and develop sustainability improvement efforts as well as tackle one of the most serious problems facing the planet, fresh water management.

Although the overall quality and value of the chapter is extremely compelling, its general structure allows readers specifically interested in certain topics to gain access to particular sections. This compartmentalized approach provides a brief introduction, main discussion, summary, and reviews for several topics ranging from recycling to the United Nations Global Compact.

Marco Ugarte, PhD, CPIM, CSCP, CQIA
The Sustainability Consortium, Arizona State University

Preface

WHY I WROTE THIS BOOK

I worked with hundreds of manufacturing companies over a thirty-year period to improve operations and supply-chain management. All of these companies made tangible reusable products, commonly referred to in the consulting profession as "little red wagons." Then I discovered the community of manufacturers who produce products that are recipe or formula based. This community, which includes the food industry, has to deal with many regulatory issues that those manufacturing "little red wagons" do not have to deal with. In addition to the differences in regulatory requirements, the food industry is undergoing rapid change and becoming ever more global and competitive. A recent special USA FDA report acknowledged the challenges those companies face. This book is written for the underserved small- to medium-size food manufacturer as a guide to world-class excellence in operations and supply-chain management.

WHAT IS COVERED IN THIS BOOK?

The book is a roadmap on how to leverage effective food safety regulatory compliance into operational and supply-chain excellence. Highlighted are the importance of cause and effect thinking, data accuracy, process simplification, process reliability and workforce development, as well as "how to" recommendations for implementing the best practices to achieve these goals. The "how to" recommendations come together in the discussions on Batch-Process ERP (Enterprise Resource Planning) and also the Lean Management System and all useful techniques within it. In the closing chapter, the rapidly developing business requirement for sustainability is covered. Sustainability has already moved for many from an optional, voluntary and "nice to do" status to a "must do" status. For the food processor that has not put together a viable sustainability plan, it soon will be imposed as a basic business requirement for doing business. Since reading by itself does not always create a high level of comprehension, each chapter has an extensive quiz, and the appendix has definitions for what for many will be a new vocabulary.

THE RESEARCH THAT WAS DONE TO WRITE THIS BOOK

For the most part, I relied on my existing knowledge developed from working with hundreds of manufacturers. I did rely on many individuals who I consider "thought leaders" to review sections of the book, which in turn often prompted the need for research, which was a valuable learning experience for me. The two topics in the book that required research were third party audit and sustainability, which proved a very satisfying experience.

ACKNOWLEDGMENTS

What I did learn in writing the book was just like the old saying that "it takes a village to raise a child," it takes peer collaboration with subject matter experts to write a reference book. I would like to thank and acknowledge the following individuals who provided considerable input in the writing of this book

- Bill Walker, the Resource Management Series Editor for CRC Press/Taylor & Francis. Not only is Bill the series editor, he is a recognized guru in supply-chain management. He provided encouragement, and also reviewed every chapter in the book and made practical recommendations.
- Wayne Labs, Senior Technical Editor, *Food Engineering Magazine.* Wayne took time out of his busy schedule, reviewed several chapters, and made recommendations based on his considerable knowledge of the food industry.
- Mike Osterling. Mike is a highly respected consultant who has helped, and continues to help, companies on their "Lean journey" to

effectively apply the techniques from the Lean toolbar, and develop a Lean progressive culture. He reviewed Chapter 7 and made extensive recommendations. Mike is also a co-author of *The Kaizen Event Planner: Achieving Rapid Improvement in Office, Service, and Technical Environments* (Productivity Press, 2007), a book highly regarded within the Lean community.

- Jerry Wright, Senior Vice President for Lean Transformation at DJO Global. Jerry is a very active thought leader within the Lean community, and DJO is an example of what is possible when a company enthusiastically and consistently embraces the "Lean journey." He reviewed Chapter 7 despite his busy schedule.
- Ron Calame is a client advisor with California Manufacturing Technology Consulting (CMTC), an organization that regularly works with small- to medium-size companies to implement Lean techniques. Ron is an old friend who reviewed Chapter 7, and made recommendations.
- Al Kuebler, CEO at Technical Impact. Al has been the CIO at many major corporations and is one of the best executives and leaders I have met in my career. He has mentored me during several periods of my career. Al reviewed Chapter 8 (sustainability) and made recommendations.
- Gerald Church, senior consultant with the CMTC. Gerald is an expert at applying Lean techniques to help companies reduce energy consumption. He developed an enhancement to value stream mapping to help pinpoint opportunities for reduction. Gerald reviewed Chapter 8 and made recommendations.
- Robert Kuhn, president at Kuhn Associates Management Advisors. Robert specializes in helping companies reduce greenhouse gas (GHG) emissions. He reviewed Chapter 8 and made recommendations.
- My wife Linda for her encouragement and support.

THE JOURNEY

The biggest challenge I encountered in writing this book was being disciplined and maintaining a reasonable pace of writing. Everyday living has many distractions.

INTENDED AUDIENCE

The intended audience includes three executives: (1) the senior executive (the CEO), (2) the executive responsible for factory operations, and (3) the executive responsible for procurement of purchased materials. On a selective basis, certain chapters are appropriate for formulation developers and quality assurance. Size of a company often separates the intended audience. Normally large companies have the resources to fund the implementation of the best practices, smaller companies less so. This book benefits both; in the case of the small- to medium-size manufacturer, it is a roadmap, for the major corporation it is a tool to help assist their supplier community.

1

Executive Overview

INTRODUCTION

> What is missing from the current food industry thinking is the linkage of regulatory compliance to operational excellence and an increased bottom line improvement.
>
> —**Wayne Labs, Senior Technical Editor,** ***Food Engineering Magazine***

The individual who will derive the greatest benefit from this book is the executive in a small- to medium-sized food or nutritional supplement manufacturing company, a segment that has been ignored by most authors who have written on how to achieve organizational excellence. The value to the reader isn't restricted to one discipline or function within a company and applies to the following:

- General management/ownership
- Regulatory compliance
- Recipe/formula management
- Planning and scheduling
- Manufacturing operations
- Materials management
- Supply chain management
- Quality management
- Distribution/logistics/transportation

The student preparing to have a career in the food manufacturing or nutraceutical industry will also benefit from reading and studying this book. Every reader will benefit because each operates as a major contributor to the process flow that results in a safe and profitable product delivered to the marketplace. Understanding the best practices of each discipline by the other disciplines creates an empathy and cooperative spirit that drives improved regulatory compliance and factory and supply chain performance. The expression "the whole is greater than the sum of its parts" is a major theme of this book (Figure 1.1).

Besides the food processor, a second beneficiary group of this book is the brand owner who uses contract manufacturers. Brand protection in the areas of quality and reliability of supply is of the highest importance, along with cost management to brand owners. This requires them to audit, inspect, coach, and nurture their contract suppliers. Information on how to improve operational performance addresses the need for brand owners to provide guidance to their contract manufacturers.

The reader will benefit from the pragmatic thinking in this book on the leveraging of regulatory compliance obligations into improved inventory

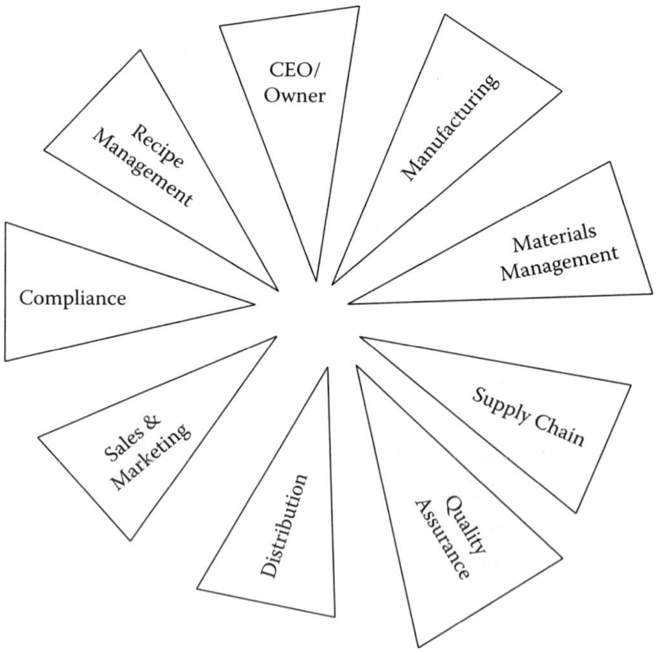

FIGURE 1.1
The whole is greater than the sum of its parts.

and supply-chain management and what proven concepts, methods, and techniques are necessary to make that happen. Think of this book as having two-for-one thinking and benefit, getting the maximum value from existing resources in a resource-constrained era. Definitions for acronyms and expressions related to concepts or techniques that support best practices can be found in the Appendix. Since this book is meant to be read in whole or in part by everyone from the CEO down the organization ladder to the factory floor supervisor, the language used is nontechnical.

Because the food and nutritional supplement industry has become significantly more complex over the last decade, supply chains are now global, and regulatory requirements have substantially increased. Much of this has been generated by consumer preferences. The consumer in industrialized nations has developed an appetite for variety and an increased focus on health and longevity. This has led to the global sourcing of many ingredients and food products from an open and free-trading world marketplace. The internationalization of the food supply chain was made possible by the end of the Cold War, which removed trade barriers and made possible offshore sourcing of products and ingredients. Consumers today expect to have food they enjoy, even if it's a seasonal food that might not otherwise be available in their geographic area. The consumer will not tolerate a rollback on the number of choices or necessarily pay a premium for this variety. From a consumer standpoint, food and supplements come from one monolithic source, the supermarket (Figure 1.2).

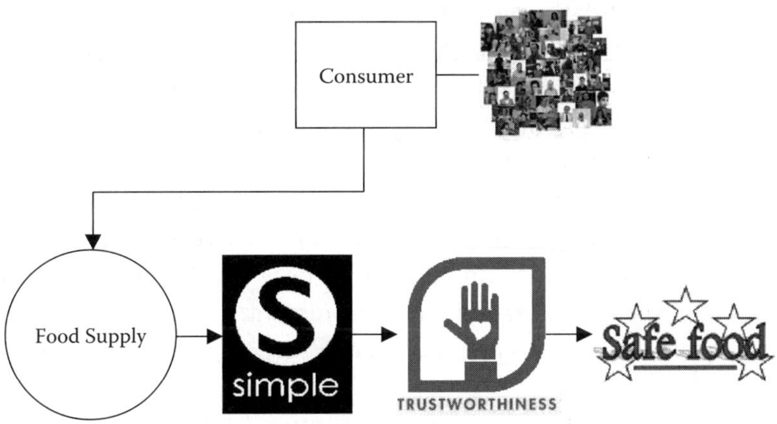

FIGURE 1.2
The consumer perception.

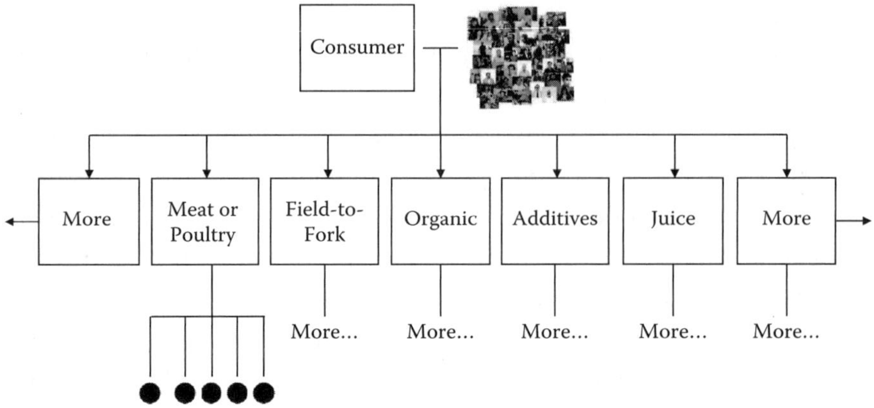

FIGURE 1.3
The food ecosystem reality.

In reality foods are produced by many micro-supply chains with different requirements, issues, and regulatory obligations. There is no definition for a micro-supply chain in our current business lexicon, and the use of this term is meant to characterize the highly specialized food products targeted to specific consumers. Food and nutritional supplements manufacturing has diversity not found in many other segments of manufacturing. For example, the automotive industry may have ultra-luxury, luxury, large mid-priced, sports, and economy cars, but there are just a few supply chains supporting it (Figure 1.3).

Since 9/11 and the seemingly endless episodes of recalls for tainted food, governments and consumers have become very concerned about the safety of the food supply chain. These concerns have resulted in more comprehensive regulatory compliance requirements imposed by national and regional governments, which necessitate a more disciplined approach to record keeping and the management of processes. While this is good from a societal standpoint, it has placed tremendous pressure on the tens of thousands of food manufacturers, many of whom are small in employee count and revenue and lack the knowledge of concepts, techniques, and supporting procedures to comply with these increased requirements. Those in the food supply chain often view these requirements as a necessary but burdensome obligation and requirement.

A few regulatory systems and obligations covered in this book are:

- Full lot traceability, all ingredients into all batches, shipped to all sources, traceability at least one up, one down, available within twenty-four hours of the request by a regulatory authority

- Hazard Analysis Critical Control Point (HACCP), a method of predetermining the steps in food processing where contamination could occur and putting safeguards in place to prevent this possibility
- Current good manufacturing practices (cGMP), guidelines that describe specific end results that minimize the possibility of corrupted food or nutritional products reaching the consumer
- Standard operating procedure (SOP), the specific internally developed procedure on how the organization will comply with a GMP obligation
- Supplier auditing, an audit of suppliers based on GMP guidelines, their SOPs, and adherence to them
- Certificates of analysis (COAs), a statement of compliance to the requirements of the ingredients and materials procured

Further complicating the situation, both food processors and nutraceutical brand owners have a large population of contract manufacturers, and the cost of entry by new competitors is relatively low.

Another consideration is the consumers' competency in the use of information technology (IT). Since they use e-mail, can access the web, use web conferencing, and have web-based professional and social networking accounts, they expect that food manufacturers will use technology to ensure the safety of the food supply chain. This expectation in turn is pushed into the food safety regulatory agencies, which apply this pressure to manufacturers. So the last decade or so has really created a difficult situation, particularly for the small- to medium-sized manufacturers.

Two other potentially challenging situations for the food manufacturing community exist: one fueled by healthcare reform and the other by global warming. In the case of healthcare reform, the emphasis will eventually shift from healthcare insurance to an individual's personal choice of lifestyle, which includes the foods consumed. The first demand has been to make manufacturers state nutritional values, but now consumers want to know what is healthy, i.e., low fat, low sodium, low carb, low calorie, organic, fiber healthy, etc. This translates into a proliferation of items produced and complexity in managing them.

Also, with obesity a growing concern in many countries, there will be pressures to increase labeling information, and with certain foods perhaps even warnings similar to that required of cigarette manufacturers and alcohol products consumed by pregnant women. Deaths related to obesity are often cited as the second leading cause of mortality in the United States, and even with the widespread use of cholesterol-reducing medications and life-saving medical interventions for obesity-related diseases, the number of deaths is still very high.

Another trend is related to global warming and an eventual "carbon tax." Many food products are heated, and in some cases refrigerated or frozen. All of these processes consume energy. Energy will be the most significant cost for many food manufacturers in the future.

Food manufacturers have experienced significant business turbulence over the last ten years, and the next ten look equally challenging. It is becoming more evident that the competitive nature and complexity of the market will increase, not decrease, in the future, requiring the implementation and adoption of best practices for inventory and supply-chain management and increasing workforce productivity on a continuous basis while increasing food safety due diligence. While many have read successful textbook case studies for inspiration on committing to a continuous improvement journey, the reality is that the high performance of these companies isn't initially necessary for competitive success. What is needed is to obtain a slight advantage and maintain it through continuous improvement.

A coworker some years ago did an informal study on the rewards of having just a slight advantage over the competition, and the resulting presentation was titled the "Principle of the Slight Advantage." What became evident was a slight advantage can generate a disproportionate reward. The research showed, for example, in the game of golf, the number of strokes per game

The Principal of the Slight Advantage and its Disproportionate Reward
PGA Tour Money Leaders – 2010

Rank	Player	Average Score	Stroke Difference Between Leader and Golfer Ranked X	Money Earned	Compensation/ Earnings Difference	Advantage PER Tenth (.1) of A Stroke
1	Ernie Els	70.2	0.4 vs. No. 3 ranked	$3,941,028	$1,088,500	+$272,125!
		70.2	1.1 vs. No. 24 ranked	$3,941,028	$2,368,393	+$215,308!
		70.2	1.2 vs. No. 45 ranked	$3,941,028	$2,929,313	+$244,109!
3	Jim Foryk	70.6		$2,852,528		
24	Adam Scott	71.3		$1,572,635		
45	Davis Love III	71.4		$1,011,715		

FIGURE 1.4
The principle of the slight advantage and its disproportionate reward. PGA Tour money leaders—2010.

between the number one prize earner and the number ten prize earner was a very modest difference, but the money earned differential was very substantial. This informal study showed that this principle carried over into the business world and many other aspects of society (Figure 1.4).

What is missing from the current food industry thinking is the linkage of regulatory compliance to operational excellence and an increased bottom line improvement. This book will link regulatory compliance to best practices in inventory, operations, supply-chain management, improved operational reliability, and Lean manufacturing. It also will discuss the eventual carbon tax and how to be proactive in addressing energy usage in factory operations.

Following is a quick overview of each chapter.

CHAPTER 2: FOOD AND NUTRITIONAL SUPPLEMENTS (NUTRACEUTICALS) INDUSTRY OVERVIEW

Chapter 2 is an overview of the food and nutritional supplement industry. Why an overview when the reader is already employed in food and nutra-manufacturing? Because the food industry is diverse and specialized, and it's possible that the reader hasn't been exposed to this industry diversity. Web-based employment recruitment forums such as the LinkedIn food discussion group really highlight this diversity and focus on specialization, as do the large number of industry associations that are listed in the appendix.

This diversity translates into many micro-supply chains within the overall food ecosystem. Many within a specific micro-supply chain might believe theirs is so unique that the best practices described and discussed in this book don't apply to them. The expression "foxhole thinking," which is the inclination to look inward and not explore better ways of operating, was coined many years ago. All the concepts, techniques, and best practices described and discussed in this book have applicability across the entire food and supplement manufacturing and delivery industry. Large food conglomerates or brand owners are more likely to understand this, but thousands of small- to medium-sized organizations may not.

What manufacturing industry outside of food manufacturing has learned over the years is that competitive benchmarking of industries outside of their specific vertical industry or style of operating has led to the discovery of significantly better ways of operating; this includes, for example, manufacturers studying retailers in the areas of customer order placement, point-of-sale data capture and instant reporting, aerospace and defense (A&D) studying automotive manufacturing to simplify device assembly, shortened lead times, and communication methods with suppliers. There are dozens of examples beyond the few mentioned.

Other points discussed in Chapter 2 are the large population of contract manufacturers and the challenges they face in getting the attention of brand owners and, once they do get their attention, continuing the relationship. On the flip side is the difficulty the brand owner has in identifying and choosing the right supplier.

The niche manufacturer of ethnic foods, for example, has challenges of being price competitive with standard nonethnic offerings and managing not to undersupply the retailer.

The global flow of finished and raw materials has increased, and in some cases the choice of suppliers has narrowed dramatically; i.e., in nutritionals manufacturing there are only a few sources of supply of certain raw materials, and they may not be of domestic origin, thereby increasing lead times and creating other complications.

Another industry-wide characteristic that is discussed is the wide range of information technology sophistication in the food supply chain ecosystem. Generally speaking, large corporations often use electronic Enterprise Resource Planning (ERP) systems that weren't designed for the processes of formulation development, the manufacture of recipe- or formula-based products, and all of the current food safety regulatory requirements. Because they have more economic and information technology staff resources than smaller organizations, they are able to connect (at great expense) other critical systems that support, for example, recipe/formulation development and food safety regulatory requirements. Medium-sized firms use selected parts of ERP supplemented by electronic spreadsheets. Small firms use electronic financial ledgers and numerous disconnected electronic spreadsheets put together to provide an isolated departmental view out of desperation driven by day-to-day business. All of these statements are generalizations, but true more often than not. The end result can be a figurative Tower of Babel within a micro-supply chain (Figure 1.5).

The final area covered in Chapter 2 is a rudimentary discussion of food safety regulatory compliance. Ongoing episodes of unintentional tainted food in the United States and elsewhere and the consequences to health and safety, as documented by the U.S. Centers for Disease Control (CDC), in combination with the reaction to the 9/11 attack in the form of the Bioterrorism

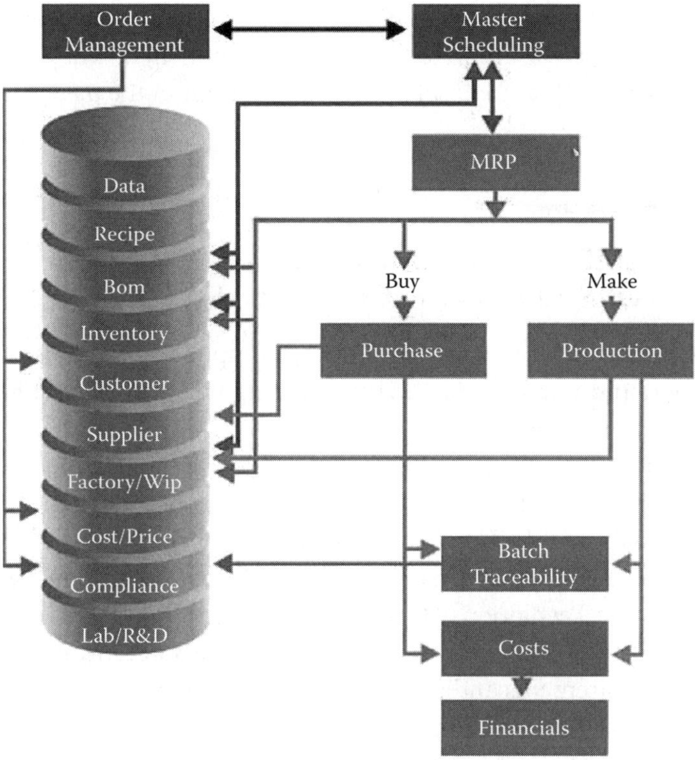

FIGURE 1.5
Process batch Enterprise Resource Planning (ERP). (From APICS International Conference & Expo, October 21–23, 2007, Denver, CO. With permission.)

Act, have elevated the concern for food safety both within the United States and internationally. The discussion of regulatory compliance obligations will be expanded in Chapter 3. Neither this brief discussion nor that in Chapter 3 is written for the food safety, regulatory compliance, or quality assurance (QA) professional, but for everyone else. The key thoughts are that the regulatory compliance regulations vary from sovereign nation to sovereign nation despite the global food economy, and some are more stringent than others. Within the United States the Food and Drug Administration (FDA), U.S. Department of Agriculture (USDA), and each of the fifty states in combination regulate and enforce food safety practices. There are other regulatory forces that are dependent on the stated declaration of the foods origin or processing environment, i.e., organic, kosher, etc.

The recent creation of a separate set of current good manufacturing practices (cGMP) for nutraceuticals apart from the food GMP will be briefly reviewed because it is reasonable to forecast (guesstimate) that some of these new requirements will migrate to food manufacturing as they represent a less ad hoc approach to food safety regulations than in the past.

CHAPTER 3: FOOD SAFETY AND REGULATORY COMPLIANCE OVERVIEW

Chapter 3 is an overview of food safety compliance for those working within food and nutraceutical manufacturing but who are not a compliance or QA professional.

The 9/11 terrorist attack on the World Trade Center was the catalyst for the Bioterrorism Act in the United States, which is best described as a top-level set of goals that the U.S. Congress directed the Secretary of Health and Welfare to enforce and make happen at a detailed level with agencies such as the FDA. A logical evolution of this act is the U.S. Food Safety Modernization Act (FSMA), and the food industry will over the next decade most likely see additional requirements and refinements to the FSMA.

Although the Bioterrorism Act itself is quite broad, the focus of our conversation is on the discipline needed to adhere to the record-keeping requirement to support lot traceability, which, given the potential gravity of a tainted food recall, is significant. Although a number of years has passed since the mandated implementation of this record-keeping requirement, the persistent marketing messages on to how achieve record-keeping compliance by ERP software developer-suppliers suggest there is significant opportunity for improvement. What constitutes an acceptable end-to-end (field-to-fork) lot trace history, and how it's compiled will change in the future, driven in part by advanced information technology. The key here is that the accuracy of those records and rapid access to them in an emergency and in subsequent discussions in this book can be leveraged into superior inventory and supply-chain management (Figure 1.6).

After 9/11, the FDA started to think in terms of analysis tools to give to the food manufacturing community; these tools simulated attacks on our food supply chains using various contaminating agents. The goal was to heighten food processors' awareness and motivate them to put in place preventive measures and recovery plans in the event of an attack. The expression "food defense" entered into the industry vocabulary and supplemented existing food safety regulations. The net outcome was a

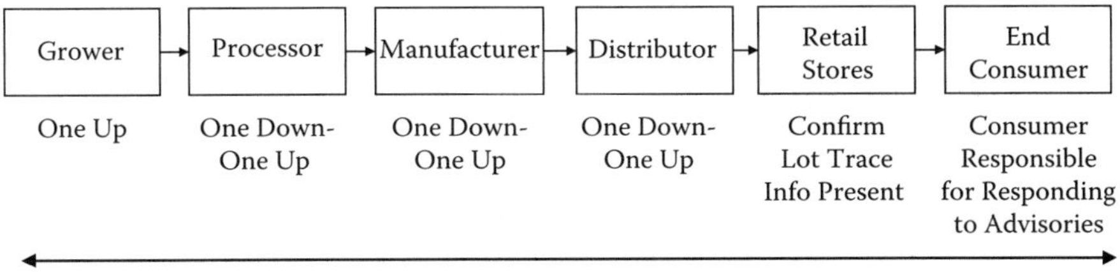

FIGURE 1.6
Lot tracking across the food supply chain. One up and one down accurate record keeping.

FDA-sponsored biotoxin attack simulation tool that will be discussed in this chapter. Also, the FDA made CARVER + Shock available, a complementary system to help both agricultural and food manufacturing companies determine how secure their facilities are and what steps are needed to take corrective action if the analysis uncovers vulnerabilities.

Another conversational thread is that historically, regulatory requirements have varied based on the type of product produced. The trend today is toward a stricter and uniform set of regulatory requirements. For example, in the past the manufacturers of nutraceuticals used the food GMP guidelines. Today nutraceutical manufacturing has a different GMP that is more rigorous than that of the food manufacturing industry. There is a slow but steady trend for the more demanding guidelines to migrate to industries with less demanding guidelines and regulatory obligations.

Many food and nutra manufacturers want to supplement their product offerings across traditional boundaries. Chapter 3 will help those focused on future business opportunities to prepare for, at a conceptual or strategic level, the associated product safety regulatory obligations. The food and the nutritional supplement industries seem to be slowly converging, and since the cGMP for nutraceuticals is recent and well thought out, it would not be surprising if the future food GMP has many aspects of it incorporated.

Hazard Analysis Critical Control Point (HACCP) for many food micro-verticals is mandatory and for almost every other is a desirable option. It is designed to address the immediate potential injury by a consumer (the public) of a food product consuming it by thinking through what processes could cause injury to them and putting in place safeguards and mechanisms to eliminate them. For organizations not obligated to comply with HACCP controls it is public declaration and manifestation of their commitment to quality and product safety when they do put these controls in place. HACCP embodies system thinking and risk mitigation. The U.S. FSMA requires that all food processors conduct a food safety

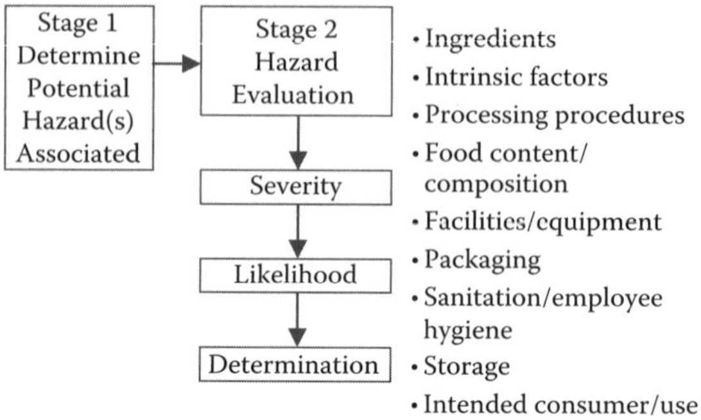

FIGURE 1.7
Hazard Analysis Critical Control Point (HACCP). USDA Appendix D: Assessing Hazard Stages.

risk assessment and put together a mitigation plan if risks are identified. Conceptually this is the front end to a full-blown HACCP. Later in this book the risk identified during the HACCP analysis will be tied to identifying the elimination of waste via value stream mapping, a highly regarded tool for those committed to the Lean enterprise (Figure 1.7).

In summary, this chapter is not directed to the regulatory compliance or quality assurance professional but those who work with them. The compliance and quality professional will be able to build on this overview and fill in the gaps with his or her coworkers.

CHAPTER 4: THE TOP TEN ISSUES IN FOOD MANUFACTURING

Chapter 4 examines an annual survey from the respected food publication *Food Engineering Magazine*. The focus of this annual survey is to identify the top ten operating issues in the food manufacturing industry. In this particular survey, the number one issue was food safety. A couple of other *Food Engineering* surveys had a similar result. All ten issues surfaced in this survey will be examined in detail. In the published survey results each issue was treated by the respondents as a stand-alone issue.

An examination of the issues clearly indicates that satisfying eight of the issues was a prerequisite to managing the number one issue of food safety. Netting it out, nine of the top ten issues were interconnected. There was a clear lack of understanding of cause-and-effect relationships. This chapter

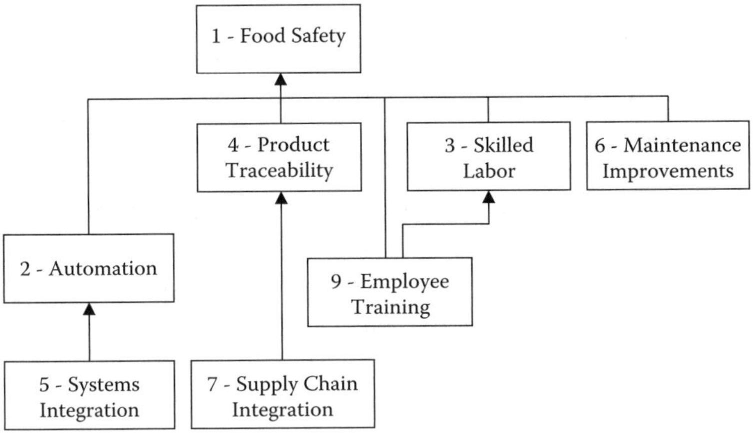

FIGURE 1.8
Industry issues. (From APICS International Conference & Expo, October 21–23, 2007, Denver, CO. With permission.)

clarifies the connectivity between the issues, thereby enabling the reader to put together an appropriate strategy and create a priority ranking on dealing with the issues, by answering the questions: Where does one start the improvement process, and what is the most effective sequence of improvements? The underlying message of Chapter 4 is that we need to think differently. What is the context? Am I addressing a symptom or a root cause? This chapter has many thought-supporting graphics because "a picture is worth a thousand words" (Figure 1.8).

CHAPTER 5: DEFINED PROCESSES, PROCEDURES, AND THE IMPORTANCE OF DATA ACCURACY

Chapter 5 describes the implied requirements of satisfying the lot traceability record-keeping obligation. It also establishes the linkage between a disciplined approach to record keeping and the management of processes and their positive relationship to superior inventory and supply-chain management.

Also discussed is the importance of well-written standard operating procedures and why HACCP thinking makes sense, even if it is not required for the products being manufactured. Following the discussion on SOP and HACCP there is a more detailed examination of the importance and organization of quarantine or hold areas, options on organizing the receiving and warehousing areas in a way that meets GMP requirements but reduces errors (mistake-proofing) and indirect costs. It also introduces a Lean thinking principle of the visual workplace.

Considerable attention is focused on data quality or data accuracy because it is one of the keys to achieving or exceeding product safety regulatory compliance and effective use of planning and control systems that support best practices and lead to improved organization performance. Questions on what needs to be accurate and why, how accurate, and how this level of data accuracy can be achieved across the organization are answered. There are a few examples that are visual and game-like that demonstrate the impact that poor data accuracy has and its implications, which in turn amplify in the reader's mind the importance of data accuracy.

CHAPTER 6: ENTERPRISE RESOURCE PLANNING (ERP) AND SUPPLY-CHAIN MANAGEMENT

In Chapter 6, how to leverage disciplined regulatory compliance into superior inventory and supply-chain management is covered in detail. It examines the techniques and proven systems that support sustainable world-class inventory and supply-chain management.

The value and potential return on investment (ROI) of utilizing Enterprise Resource Planning (ERP) is discussed using a study by two respected organizations, one a university and the other a professional association. Also examined are how ERP has evolved over decades of trial and error and what business purpose each of its subsystems serves. The difference between traditional ERP versus process batch ERP that supports recipe-based products is highlighted and explained. How data accuracy and discipline in executing business processes affect the effective use of ERP is covered and emphasized. The techniques within ERP that help support achieving data accuracy and a disciplined approach to status reporting are also covered.

Chapter 6 concludes with a focus on managing a company's supply chain, the reasons behind the rapid growth of out-sourcing, which have often created complex supply chains, and the issues that have evolved over the last decade. Included are recommendations on who a corporation can work with to improve performance in both participating in a supply chain as a supplier and managing "upstream" suppliers as a customer.

Another thread in the chapter is a discussion on the demand-driven supply chain that incorporates the Lean operating philosophy. Lean practitioners often call this the push-versus-pull debate. Also discussed are the current limitations of computer-based planning and control systems highlighted in the *Wall Street Journal* article titled "Clarity Missing in Supply Chains." The article is a case study that elaborated on the devastating impact the lack of a systematic method of communicating across the supply chain had on a global supply chain. Why this happened will be examined and how it can be avoided in the future outlined.

CHAPTER 7: HOW TO ORGANIZE THE WORKPLACE—LEAN THINKING

The compatibility in food and nutra manufacturing with the Lean manufacturing philosophy and its proven techniques are examined in this chapter, including techniques such as 6S (5S), value stream mapping, A3, kaizen events, the compatibility of HACCP and value stream mapping, and the importance of standard work.

CHAPTER 8: SUSTAINABILITY

The final chapter examines both energy and water usage management, and the need for strategic thinking and action, and its probable impact on both food and supplement manufacturers.

Carbon

Some form of a direct or indirect carbon tax on energy usage in the future has been predicted. Energy from carbon-based raw materials has gotten very expensive over the last decade, and is forecasted to become even more expensive. The cheap energy era is over. In response, methodologies to account for and create plans to reduce carbon-based energy usage have been developed. Major corporations now use these methodologies and publicly report progress in a fashion similar to that for financial statements from publicly traded corporations. Over eighty-five percent of the Global 500 report their emissions status and targets. How well a corporation manages and reduces its carbon emissions is now synonymous with the management of risk. Most large food manufacturing conglomerates and food retailers have been focused on reducing carbon emissions within their corporate boundaries, and now are starting to focus on their supply chains.

This chapter provides an overview of the *de facto* frameworks, methodologies, and data repositories for reporting and managing greenhouse gas (GHG) emissions, including the GHG Protocol, ISO standards, the Carbon Disclosure Project, and the UN Global Compact.

Water

Water supplies are also a concern in many parts of the world, and in those areas the cost of supply will continue to increase. Recurring droughts in the southwest region of the United States are an example of where the mandated reductions in use have actually caused higher prices for water to cover water system overhead costs. Australia has taken the lead and invested in massive water desalination capability, and others will follow suit. Food products are dependent on water availability and reliable cost-effective sources of supply for future decades, if not beyond. In recognition of this water crisis the UN added a special program to its Global Compact, the CEO Water Mandate.

The organization most vulnerable to increased energy and perhaps water costs is the contract manufacturer who heats or cools or freezes a product and competes on price. Does the manufacturer really have costing systems in place to understand what costs it incurs to make a batch of product A, and is it confident it can compete on price and genuinely understand the real costs and make money?

In response to this crisis a comprehensive methodology to manage and reduce water consumption has been developed to support efforts within the four walls of the enterprise and the supply chains feeding it. The methodology is the Water Footprint Network, which will be examined in this chapter.

With major corporations/customers now starting to focus on having their suppliers reduce carbon emissions, and water usage, the role of the Supply Chain Operations Reference (SCOR) supply chain modeling framework, discussed in Chapter 7, will be revisited. Although SCOR started as a method to profitably utilize a supply chain, it can serve another concurrent purpose: provide a foundation for the GHG and water management methodologies to piggyback on.

This is the decade that sustainability becomes a mandated requirement because it has economic risk; getting started today will make the task of implementing appropriate measures easier.

APPENDIX

The book ends with an appendix that is information rich and includes:

- Definitions
- Quiz answers
- References
- Relevant website addresses
- Recommended reading

THE READER ROI

Things the reader will learn from this book:

1. The reader will have a "how to" roadmap on leveraging his or her existing investment in food safety regulatory compliance into superior inventory, operations, and supply-chain management.
2. The reader will be able to articulate how Lean operating principles and the discipline required of regulatory compliance support and reinforce each other.
3. The reader will understand what needs to be improved, in what sequence improvements must be addressed, how one improvement feeds another, and the prerequisites for each improvement.
4. The reader will know how to leverage cost-effective information technology.

Some tips for the reader are to create a mental framework or context of the major section or chapter before reading it by reading the first sentence of each paragraph and then reading the entire section or chapter. This technique can be helpful to digest the information more fully. Also, at the conclusion of each chapter or logical subsection there are mini-quizzes. These have two purposes: to let the reader do a private test of what he or she has comprehended, and to give the executive or supervisor talking points for team review.

QUIZ

1.1. This book will benefit only the manufacturer of food products within a specific supply chain.
☐ True ☐ False

1.2. This book will benefit the manufacturers of food products or nutraceuticals irrespective of the supply chain they participate in.
☐ True ☐ False

1.3. This book provides no value to the brand owner who utilizes contract manufacturers.
☐ True ☐ False

1.4. The manufacturer of food and nutra products is a process flow requiring the participation of many departments and disciplines to achieve food safety regulatory compliance, operational efficiency, and superior supply-chain management.
☐ True ☐ False

1.5. The targeted reader for this book is restricted to an organization's CEO.
☐ True ☐ False

1.6. Supply chains will become less complex because of consumer preferences.
☐ True ☐ False

1.7. There will be a reduction in food safety compliance requirements in the future.
☐ True ☐ False

1.8. The discipline required for product safety in terms of accurate record keeping and data reporting accuracy is an economic burden with no payback.
☐ True ☐ False

1.9. The following chapters of this book are intended to provide guidance on leveraging food safety excellence into superior business operations.
☐ True ☐ False

1.10. There is a high probability that the U.S. healthcare reform will lead to changes in the product portfolio and product labeling.
☐ True ☐ False

1.11. Future energy costs for those who heat, cold store, or freeze a product will not be a major consideration.
☐ True ☐ False

The answers to these questions can be found in the Appendix, as well as definitions, website addresses, and recommended reading.

2

Food and Nutritional Supplements (Nutraceuticals) Industry Overview

This chapter is an overview of the food and nutritional supplement industry. Why an overview for readers already employed in the food industry? Because both the food and nutritional supplement industry are diverse, exceptionally large, and specialized; the different permutations of even a simple food product are staggering. The USDA food "plate" (formerly the pyramid) is simple (Figure 2.1):

- Grains
- Vegetables
- Fruits
- Milk
- Meats and beans

But from this straightforward food plate there are an almost infinite number of food products from a huge large manufacturing base, and there are different forms of each standard product, for example, standard, organic, kosher, fresh, dry, frozen, canned, prepackaged meals, low sodium, and cultural taste adjustments to deal with ethnic micro-markets. The natural characteristics of the food product being processed can influence the processing and skills of the manufacturing staff, for example, acidic ingredients. If you were a regulatory enforcement professional, life wouldn't be easy.

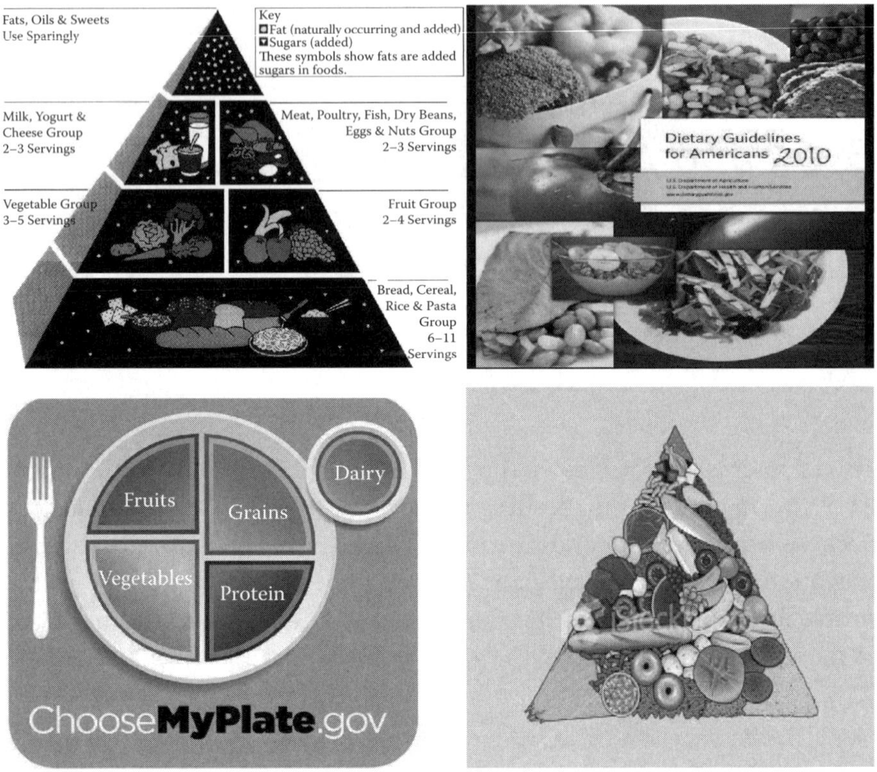

FIGURE 2.1
Four different views of the recommended nutritional plate.

THE SIZE AND COMPLEXITY OF THE FOOD INDUSTRY

Let's examine the size of the food and not the nutra manufacturing base. The numbers are from the U.S. Food and Drug Administration (FDA), and the true worldwide number is many times larger. The FDA states:

> FDA registration is required for all companies that manufacture, process, pack, or store food or beverages that may be consumed in the United States. Companies located outside the U.S. must designate a U.S. Agent for FDA communications.

Estimates from the FDA are that the total number of food facilities that must register with the FDA is approximately 420,000, approximately half of which are domestic. The numbers are truly staggering!

THE FOOD INDUSTRY IS INCREASINGLY GLOBAL AND INTERCONNECTED

The global flow of finished food products and ingredient raw materials has increased, and in some cases the choice of suppliers has narrowed dramatically; for instance, in nutritionals manufacturing there are only a few sources of supply of certain raw materials, and they may not be of domestic origin, thereby increasing lead times and creating other complications.

A special report published by the FDA, Pathway to Global Product Safety and Quality, examines the tremendous increase in global trade of regulated products. While this report covers all of the products the FDA has the responsibility to regulate, it also extensively covers the food and beverage industry and its globalization. The report states:

> The manufacturers and producers that FDA regulates face intense pressure to lower costs and improve productivity, fueling a cycle in which the quest for efficiency leads to increased production abroad and higher volumes of imported products to regulate. Goods entering the U.S. will come from new and different markets, flowing through long, multistep processes to convert globally-sourced materials into finished goods.

The report goes on to forecast:

> The cumulative effect of these trends will ensure that 10 years in the future, the world will be very different than it is today, with a dramatic increase in the global flow of goods, including increases in imports to the U.S.

What caused the rapid globalization of the food supply chain? There are a number of contributors to this food globalization trend:

- The end of the Cold War. The world went from three distinct trading communities—(1) the West, (2) the Soviet block, and (3) nonaligned countries—to one essentially nonideological world zone.
- The World Trade Organization (WTO) and the general acceptance by the public that world trade is good.
- The demand from the countries with an advanced economy for more variety in foods, acceptance of foods originating outside their country, and a preference for specific foods irrespective of traditional seasonal availability.

- The international agreements that subsidized global trading, i.e., the nontaxation of fuel used in global export/import transactions.
- Advanced and cost-effective information technology has made possible 24 (hours)/7 (days) communication. The trading partner no longer needs to be local.
- Food science has made many advances in extending food product freshness and preservation.

The report's recommendation for those producing regulated products in the advanced industrialized world is stated in a section titled "The Productivity Imperative":

> Emerging markets are riding a virtuous [*sic*] growth cycle, propelled by larger and younger working populations. In the rich nations of the developed world, by contrast, low birth rates and graying workforces will make it enormously difficult to maintain what economist Adam Smith called "the natural progress of opulence." These countries' best hope for keeping the wealth creation engine stoked is *improved productivity—producing more with fewer resources.* [Italics added]

What might partially impact this global flow of food, at least in theory, is taxing fuel used in international trade. Currently there is no tax on fuel used by ocean freighters and air freight for international transport of goods, including food (see Appendix). Since the concept of a carbon tax is slowly gaining acceptance to combat global warming, taxation of fuels used in international trade is a possibility since the amount of fuel, and hence hydrocarbon generation, is significant. If this does happen, the use of the total landed cost (TLC) costing methodology will be more widely adopted on imported foods and ingredients with low profit margins.

THE FOOD INDUSTRY'S EVER-GROWING PRODUCT PORTFOLIO

The different forms of a food product add to the complexity of the overall food manufacturing ecosystem:

- Field harvested and uncooked
- Organic

- Antibiotic-free
- Religious considerations
- How the food was prepared based on food chemistry
- Seasonal quotas (fish)
- A specialized food ingredient, i.e., seasonings

Another industry dynamic is product proliferation. Consumer demands, in combination with the modularity of a recipe-based product, continue to drive new product offerings creating product portfolio complexity. This portfolio complexity will require tough decisions about the impact of a new product on existing products and a need to understand when to retire a product from the portfolio. A simple example of product modularity and the freedom it brings to business is best illustrated by a simple luncheon menu in which the consumer is offered only five category choices that can result in a very large number of end-item food products. The example menu in Figure 2.2 is both an illustration and a quiz. How many end items can be produced from this menu? The answer is in the Appendix.

PROTEIN CHOICES

☐BEEF PATTY
☐ROAST BEEF
☐CHICKEN SALAD
☐PASTRAMI (HOT)
☐SALAMI
☐CORNED BEEF (HOT)
☐HAM
☐HAM & SWISS
☐CHEESE
☐TURKEY (SLICED)
☐REUBEN
☐SHRIMP SALAD
☐TUNA
☐ROAST PORK
☐EGG SALAD

BREAD CHOICES

☐WHEAT
☐WHITE
☐RYE
☐PUMPERNICKEL
☐ROLL
☐SUB

CONDIMENTS

☐MUSTARD
☐MAYONNAISE
☐BUTTER
☐LETTUCE
☐TOMATO
☐ NONE

SALADS

☐CARROT
☐GERMAN POTATO
☐MACARONI
☐COLESLAW
☐POTATO
☐NONE
☐TOSSED GREEN
☐HOUSE SPECIAL
☐GREEK

BEVERAGES

☐TEA
☐PEPSI
☐COKE
☐DIET PEPSI
☐DIET COKE
☐DR. PEPPER
☐VERNOR'S GINGER ALE
☐COFFEE
☐MILK
☐ICED TEA
☐NONE

NUMBER OF OPTIONS

____

NUMBER OF CONFIGURABLE LUNCHES

____

FIGURE 2.2
Typical customer-configurable luncheon menu.

There is an underlying body of knowledge (BOK) on product modularity, and for those unfamiliar with it, it is recommended that they read *Mass Customization: The New Frontier in Business Competition* by B. Joseph Pine. This book is the definitive text on product modularity, which seems to be applicable to almost every type of product made: computer, car, recipe-based product, etc.

While there doesn't seem to one authoritative source of information on how many different food products are currently offered in each region of the world, there are informal estimates that there are hundreds of thousands of products offered in the United States alone. For both a food or supplement manufacturer, managing an ever-changing product portfolio is challenging and requires integrated recipe/formula management, supply planning, operations and inventory management, quality management, record keeping, and distribution systems. Also required is a workforce educated on the best practices for each discipline supporting the supply chain processes.

Food safety regulatory agencies have managed this huge number of food products by creating guidelines and approved lists of ingredients using ingredient history and well-thought-out definitions, separating ingredients of a dietary supplement out from accepted food ingredients, understanding the impacts certain foods have on our diverse population, creating rules for these subset groups, and establishing guidelines in certain areas on industry self-regulation.

PACKAGING—ANOTHER CHALLENGE

A persistent complication for food and nutritional supplements product manufacturers is packaging and labeling. The concern for the impact of packaging on food safety seems always to be in flux. What was viewed as a major breakthrough in packaging for food preservation and consumer safety seems to be revisited periodically just when everyone believes a problem has been resolved. The increase in research technology capabilities has made this revisiting possible, which is great from a societal standpoint, but difficult for the manufacturer since packaging design and equipment are often expensive. Also, the importance of packaging is reflected in the U.S. Bioterrorism Act, which requires the lot tracing of packaging materials that have direct physical contact with the consumer product.

Another management complexity related to packaging is the consumers' demand to not only have the food in different configurations, i.e., low sodium, organic, etc., but to have it packaged to suit an individual, a standard, and a family-size portion. This adds to the number of items that must be planned and managed, as well as the resulting complexity.

CONTRACT MANUFACTURING

Another characteristic of both industries is the large population of contract manufacturers. These companies often have additional challenges of having auditing not only by food safety regulatory authorities but also by their customer, the brand owner. On the flip side is the difficulty the brand owner has in identifying and choosing the right supplier.

ON THE HORIZON

The success of genetic engineering in the food and agriculture industries has created enthusiasm within the food industry, which has invested billions of dollars into nanotechnology research. This is the new frontier for both the food and nutritional supplement industries. One study on nanotechnology states:

> Keeping leadership in food and food processing industry, you have to work with nanotechnology and nano-bio-info in the future. The markets are changing already. The future belongs to new products, new processes with the goal to customize and personalize the products. (Helmut Kaiser Consultancy Group)

This study also predicts that nanotechnology will be used in 40% of the food industries by 2015; please see the Appendix for the website link for the overview of this study.

The three areas in which nanotechnology seems most relevant are food fortification and modification, interactive "smart" food and smart packaging, and food tracking.

It appears that the acceptance of nano-food products isn't uniform around the world, with countries such as the People's Republic of China being an eager early adopter, while the United States is more slowly coming on board.

Currently there are no laws governing the use of nano-materials in consumer products to ensure that they do not cause harm to the public using them, the workers producing them, or the environmental systems into which waste nano-products are released. This doesn't mean that regulatory authorities are ignoring the issue. The U.S. FDA Nanotechnology Taskforce (2007) states:

> The FDA has not established its own formal definition, though the agency participated in the development of the NNI definition of "nanotechnology." Using that definition, nanotechnology relevant to the FDA might include research and technology development that both satisfies the NNI definition and relates to a product regulated by FDA. [See website addresses in the Appendix.]
>
> Because FDA regulates products based on their statutory classification rather than the technology they employ, FDA's regulatory consideration of an application involving a nanotechnology product may not occur until well after the initial development of that nanotechnology.

Some people believe that the genetic engineering of crops and exploration of nanotechnology by industry have been contributors to the movement to organic foods and supplements. The net result of all this is that the product portfolios for most manufacturers will increase, bringing complexity, including the need for good information tools, comprehensive recipe/formula management, regulatory compliance, and quality assurance, as well as integrated planning and control systems. Imagine if you made chicken soup and needed to cover most of the consumer preferences, such as a traditional recipe (comfort food?), a low-sodium version, a low-calorie version, an organic version, and a nanotechnology-enhanced version.

FOOD SAFETY

If a company exports products, it has to deal with the product safety requirements of each country exported to, and the company's product safety obligations in its home country, which may be different than those of the country importing the product. Global standards such as ISO 22000 are an attempt to have a unified global standard.

The type of product and location can radically change the food safety authorities with which a company must collaborate. Using the United States, for example, a company could conceivably work concurrently with the FDA, USDA, and state, county (province), and city inspectors, and finally, the brand owner if the company is a contract manufacturer. If there is an allergen ingredient in the food product, it adds to the complexity: Does the manufacturer state that the facility processes ingredients that are listed as allergens and potentially lose market share or build a separate and sealed-off facility?

AN EXAMPLE OF FOOD INDUSTRY SELF-REGULATION

In the United States, an example of industry self-regulation is the ingredient/additive approved-for-use list known as the generally recognized as safe (GRAS) classification. GRAS began in the late 1950s when all ingredients (additives) that were generally accepted by knowledgeable experts as safe and had a history of being so were "grandfathered in" to the then new GRAS list.

This phase of GRAS was followed by a period of submittals by manufacturers to the U.S. FDA for approval of new ingredients not listed as GRAS. It appears (speculation) because of the workload, the FDA created guidelines and let manufacturers self-certify with no FDA premarket review. The FDA website states:

> FDA has several lists of GRAS substances. Importantly, these lists are not all-inclusive. Because the use of a GRAS substance is not subject to premarket review and approval by FDA, it is impracticable to list all substances that are used in food on the basis of the GRAS provision.

Not only is the product portfolio becoming more complex to manage, but so is the recipe or formula development relative to the use of new nonapproved GRAS ingredients and the processes to get new ones listed. Although the FDA has made the GRAS new ingredient process simpler, the placement of responsibility for due diligence on the manufacturer hasn't made the challenge of managing an ever-increasing product portfolio easier.

THE INFORMATION TECHNOLOGY CHALLENGE

Another industrywide characteristic that is discussed is the wide range of information technology sophistication in the food supply chain ecosystem. Generally speaking, large corporations often use electronic Enterprise Resource Planning (ERP) systems, which weren't designed for recipe/formula-based products, recipe management systems, forecasting systems and demand management systems, bar coding systems within the factory, so are often at a suboptimum level. Medium-sized firms use selected parts of ERP supplemented by electronic spreadsheets. Small firms use electronic financial ledgers and numerous disconnected electronic spreadsheets put together to provide an isolated departmental view out of desperation driven by day-to-day business. The result of this inconsistency is poor quality (accuracy) of information, which not only affects food safety record keeping but also hampers management decision making. Since every enterprise is a member of a supply chain, either as the supply chain master (end-branded product customer) or as a supplier, this inconsistency gives meaning to the reworked expression "a supply chain is as strong as its weakest link." These statements are generalizations but true more often than not. This data quality issue, in combination with current deficiencies in the body of knowledge on supply-chain management today, which is discussed in Chapter 7, has figuratively created a Tower of Babel within many micro-supply chains (Figure 2.3).

LEARNING FROM OTHERS—THINKING OUTSIDE THE BOX

Many manufacturers within a specific micro-supply chain might believe theirs is so unique that the best practices described and discussed in the following chapters of this book don't apply to them. The expression "foxhole thinking," which is the inclination to look inward and not explore better ways of operating, describes this type of thinking. All the concepts, techniques, best practices discussed in this book apply across the entire food and supplement manufacturing and delivery industry. Large food conglomerates or brand owners are more likely to understand this, but thousands of small- to medium-sized organizations might not.

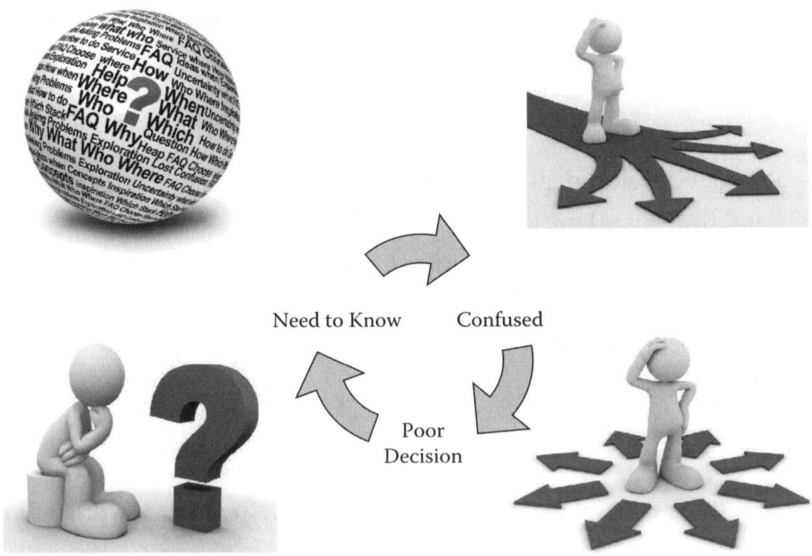

FIGURE 2.3
The end result of poor data quality and lack of understanding best practices.

What the manufacturing industry outside of food manufacturing has learned over the years is that competitive benchmarking of industries outside of their specific vertical industry or style of operating has led to the discovery of significantly better ways of operating. This includes, for example, manufacturers studying retailers in the areas of customer order placement, point-of-sale data capture, instant reporting, and aerospace and defense (A&D), and studying automotive manufacturing to simplify device assembly, shorten lead times, and improve communication methods with suppliers. There are dozens of examples beyond the few mentioned.

RESOURCES TO ASSIST THE FOOD INDUSTRY

Another indicator of the food and supplements industry diversity is reflected in a nonprofit professional association, the Institute of Food Technologists (IFT). IFT has thousands of members and is dedicated to the advancement of food science and technology. It has been serving food science professionals and the companies for whom they work for more than seventy years, and the scope of educational conferences covering all aspects of food science and ad hoc education and topic-focused forums clustered in the following eight areas:

- Food safety and defense
- Product development (formulation)
- Food health and nutrition
- Emerging technology and ingredient innovations
- Public policy and regulations
- Education and professional development
- Food processing and packaging
- Sustainability

IFT's listing of recognized universities in North America featuring a food technology curriculum is another indicator of the multiplicity of this industry and the need for skilled professionals to staff manufacturing organizations:

Alabama A&M University
University of Arkansas
Brigham Young University
University of California–Davis
Cornell University
University of Florida
University of Guelph
University of Illinois
Kansas State University
Louisiana State University
University of Manitoba
University of Massachusetts
Michigan State University
Mississippi State University
Instituto Technologico y de Estudios
North Carolina State University
North Dakota State University
Ohio State University
Pennsylvania State University
Rutgers—The State University
Texas A&M University
Utah State University
Washington State University/University of Idaho

University of Alberta
Auburn University
University of British Columbia
California Polytechnic State University
University of Delaware
University of Georgia
University of Idaho/Washington State
Iowa State University
University of Kentucky
University of Maine
University of Maryland
McGill University
University of Minnesota
University of Missouri
University of Nebraska
Oregon State University
Purdue University
San Jose State University
University of Tennessee
Virginia Tech
University of Wisconsin–Madison

There are still many small colleges with a food technology curriculum and approximately more than two hundred focused industry trade associations and special interest forums in North America alone.

Web-based employment recruitment forums such as the LinkedIn food discussion group really highlight this diversity and focus on specialization, as do the large number of industry associations listed in the Appendix of this book. A food professional whose company makes dairy-based products might not fully understand the issues of a company making grain-based bread or a field-to-fork or meat product. The probability that everyone who reads this book understands all the requirements across the entire spectrum of food and supplements manufacturing is unlikely because of the degree of complication (Figure 2.3).

SUSTAINABILITY—ONCE OPTIONAL, SOON REQUIRED

Sustainability and social responsibility are an obligation that is growing in importance among respected manufacturers who are "thought leaders" within their industries. On a global level this is reflected in the number of corporations committing to the United Nations Global Compact, which is the world's largest corporate citizenship and sustainability initiative.

The UN Global Compact is a strategic policy initiative for businesses that are committed to aligning their operations and strategies with ten universally accepted principles in the areas of human rights, labor, environment, and anticorruption. These ten principles are grouped into four categories:

- Human rights
- Labor standards
- Environment (sustainability)
- Anticorruption

The importance of each issue can be understood by examining the labor issues once faced by Nike, the manufacturer of athletic shoes and apparel. Many years ago Nike faced considerable criticism for its treatment of offshore workers. To Nike's credit, it responded by taking action to correct this situation and also moved aggressively to remove toxic materials from all its products, many of which are difficult to detect by the consumer. At the time Nike made this commitment to removing toxic materials there was no overt demand for it to do so. Good citizenship often goes unnoticed. While this may not appear to apply to food manufacturing or nutritional supplements manufacturing, it actually does; the current food supply chain is global.

With this global network come variations, sometimes extreme, in the treatment of workers, legal fairness, and respect for the ecosystem (sustainability). Compliance with the principles of the compact translates into obligations incumbent on the global brand owner or those who use contract manufacturers or those who purchase raw materials from offshore suppliers to conform to a global standard. The UN Global Compact looks like it is another business commitment and consideration that can't be ignored.

Below is a partial list of companies recognized in North America that have made a commitment to the compact:

Campbell Soup Company
Deans Beans Organic
General Mills
Molson Coors Brewery
PepsiCo, Inc.
Talier Trading Group
Exportalizas Mexicacas, S.A. de C.V.
Grupo Industrial Lala, S.A. de C.V.
Industria Envasadora de Queretaro, S.A. de C.V.
Coca-Cola Enterprises
E Covita
Green Mountain Coffee
Monsanto
Starbucks Coffee Company
Coca-Cola Company
Fomento Economico Mexicano, S.A. de C.V. (FEMSA)
Grupo Nestle
Sesajal, S.A. de C.V.

The UN Global Compact document "Supply Chain Sustainability: A Practical Guide for Continuous Improvement" has in it the following definition:

> Supply chain sustainability is the management of environmental, social, and economic impacts and the encouragement of good governance practices throughout the lifecycles of goods and services.

It also states:

> Many companies' most significant impacts on the environment and society occur in their supply chains... supply chains consist of continuously evolving markets and relations.

The report also addresses the "why" question:

> The objective of supply chain sustainability is to create, protect and grow long term environmental, social and economic value for all stakeholders involved in bringing products and services to market.

While many of the companies mentioned are very large in terms of revenue, the document clearly states that the intent is to "extend sustainability to small and medium sized enterprises."

A quick review of the food companies who have committed to the UN Global Compact indicates very few food suppliers from the People's Republic of China (PRC) have endorsed this path. This might be significant in that the PRC is the third largest exporter of food products to the United States. Also, since there isn't a separate category for nutritional supplements and India is a major source of supply for botanicals used in advanced formulations, and there are very few Indian food companies who have committed to the UN Global Compact, it might be prudent for a supply chain orchestrator (customer) to put safeguards in place, particularly for the PRC.

Those who have also joined the movement to the compact are professional associations such as the Association for Operations and Supply Chain Management (APICS). APICS has a history of codifying a body of knowledge, creating courseware to disseminate it, and has created certifications to confirm competence in understanding and applying it.

In summary, the UN Global Compact will become progressively more pervasive, and both customers and suppliers will be prudent to add this additional layer of discipline to their other efforts in food safety and food defense.

SUMMARY

The purpose of this chapter was to give a high-level overview of the food and nutritional supplement industry. Why an overview for those readers already employed in the food industry? Because both the food and nutritional supplement industry are diverse, exceptionally large, specialized, and the different permutations of even a simple food are staggering.

QUIZ

2.1. Estimates from the U.S. FDA are that the total number of food facilities that must register with the FDA is approximately 100,000.
☐ True ☐ False

2.2. The food safety regulatory compliance treatment by food safety regulators is uniform across all food and nutritional supplement manufacturing.
☐ True ☐ False

2.3. Current consumer demands for more variety and dietary considerations are increasing product portfolios and complexity.
☐ True ☐ False

2.4. There is no authoritative text on understanding product modularity and simplifying portfolio management.
☐ True ☐ False

2.5. Supply chains have become more global.
☐ True ☐ False

2.6. Fuels used in international trade, both ocean and air transport, are taxed.
☐ True ☐ False

2.7. The U.S. FDA in the case of generally recognized as safe (GRAS) submissions has put the burden of proof on manufacturers.
☐ True ☐ False

2.8. Nanotechnology will have no impact on the food and supplements industries.
☐ True ☐ False

2.9. Contract manufacturers are not widely used within the food industry.
☐ True ☐ False

2.10. The effective utilization of information technology for planning, control, data integrity, and record keeping is a constant among small, medium, and large supply chain participants.
☐ True ☐ False

2.11. Intra-industry benchmarking is common within food and supplement manufacturing.
☐ True ☐ False

2.12. Food and supplement manufacturing has limited support from educational institutions and forums.
☐ True ☐ False

2.13. There is limited commitment to the United Nations (UN) Global Compact, including the sustainability component.
☐ True ☐ False

2.14. All nations involved in the global food supply chain have demonstrated enthusiasm for the UN Global Compact.
☐ True ☐ False

The answers to these questions can be found in the Appendix along with definitions, website addresses, and recommended reading.

3

Regulatory Compliance Overview

This chapter is directed at those in manufacturing who are not food safety and regulatory compliance professionals. It's an overview that should fill in a few gaps in the readers' knowledge. The purpose of this chapter is to show the connection between rigorous compliance to food safety regulations and excellence in operations and supply-chain management, which translates, when acted on with understanding and commitment, into increased company profits and job security for employees.

Most, if not all, food and nutritional supplement manufacturers employ trained food safety and regulatory compliance specialists along with food scientists. But the obligation for food safety is shared by every employee, starting with the CEO, and continuing with managers, supervisors, and workers, but not just these specialists. It is an organizational and societal obligation for everyone.

The need for food safety is highlighted by the U.S. Centers for Disease Control (CDC), which has reported more than five thousand deaths and hundreds of thousands of individuals hospitalized consistently on an annual basis because of tainted food. The global numbers are much larger.

This suggests that the battle to achieve food safety isn't easy and requires the involvement of each employee. The various international, federal, state/province, and local governments have food safety guidelines and enforcement capabilities. The three most well-known federal agencies in the United States are:

- Food and Drug Administration (FDA), which has oversight responsibilities beyond food manufacturing, and also includes nutritional supplements, pharmaceuticals, cosmetics, medical devices, etc.

- U.S. Department of Agriculture (USDA) via its Food Safety and Inspection Service (FSIS), which is chartered to ensure the safety of meat, poultry, and egg products.
- U.S. Environmental Protection Agency (EPA), which is responsible for the safety of drinking water as mandated in the Safe Drinking Water Act (SDWA). It also is chartered with determining the toxicity of pesticides and their use in agriculture.

Beyond these three there are twelve other agencies in the United States involved in food safety, which has spurred discussions about rationalizing and integrating food safety oversight. The responsibilities and cooperation among the three most well-known agencies in a supply chain context and who they regulate in the supply chain are depicted in Figure 3.1.

There are critics as to the performance of these regulatory organizations in protecting the public food supply, and some argue for an integrated food safety agency—unlike the current separation of oversight—and also because the FDA, USDA, and EPA have additional responsibilities; e.g., the FDA is responsible for the pharmaceutical approval processes and the guidelines for the manufacturing of them. In response, these regulatory organizations often ask for increased legal authority, funding, and workforce.

Long-standing organizations such as NSF International provide a range of services and often act as a liaison between regulatory authorities and manufacturers. NSF International is a not-for-profit, nongovernmental organization, a world leader in standards development, product

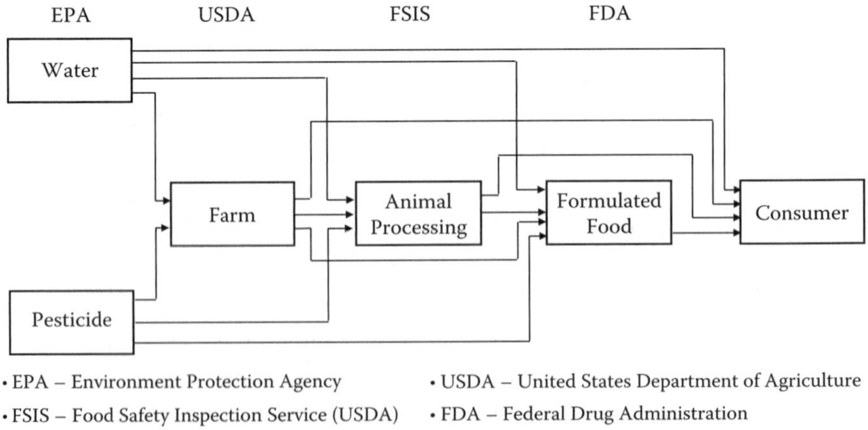

FIGURE 3.1
The U.S. food safety regulatory supply chain: Simplified view.

certification, education, and risk management for public health and safety. Its expertise and assistance also covers workplace safety and environmental sustainability. NSF has earned the Collaborating Center designations by the World Health Organization (WHO) for Food and Water Safety and Indoor Environment. In all likelihood NSF is a resource to your company's food safety and regulatory compliance specialists.

Since both the food and beverage and nutraceutical industries consist of many specialized products serving "micro-vertical" markets, there are numerous highly focused advocacy associations that lobby for their member companies and offer forums and educational offerings. Again, the regulatory compliance specialist individual(s) at your company may also be involved with the appropriately focused association(s).

If your company operates as a contract manufacturer, there is another layer of scrutiny and audit, this time from the customer. Many customers retain third-party food safety audit firms that are highly specialized, experienced, and objective in their evaluations, and all of them interact with the food safety and compliance specialist(s) at your company.

There can be other regulatory requirements if your company exports or manufactures nutraceuticals (nutritional supplements) or cosmetics, or offers products defined as organic or kosher. This is not unusual in today's competitive marketplace as companies strive for higher-profit-margin products. A few complications your food safety/regulatory compliance specialist has to deal with include:

- Exports: The food safety regulations in major international markets aren't always the same for many products. An example: The requirements in Japan for product X might be more stringent than those in the United States and Europe, and the differences can be in any combination of trading export/import relationships. With regard to the United States, Europe, Japan, and other markets, there are attempts to harmonize these differences, an example being ISO 22000.
- Manufacturing "other" products: The good manufacturing practices (GMP) requirements for food, nutraceuticals, and cosmetics are different. The majority of requirements for all three are similar, but there are distinct differences. Nutraceuticals have the most demanding of the three, but over time this could change and probably will!
- Organic: This is now in the mainstream and has its own set of compliance requirements above and beyond the normal obligations of industrialized food production.

For those managing manufacturing operations or their companies' supply chain, and who aren't specialists, it makes sense to understand the food safety regulations, why they were created, and the concepts they are built on. The rest of this chapter covers:

- Pre-9/11 best practices and regulatory compliance obligations
- The Bioterrorism Act—a nontechnical overview, and where lot traceability is required in the food supply chain
- Food safety outside the United States
- Going beyond the Bioterrorism Act—the FDA-sponsored biotoxin attack simulation capability
- Results and developments since the Bioterrorism Act

PRE-9/11

Food safety and regulatory compliance has been with us both formally and informally for many years. 9/11 forced a formalizing of some of these requirements and extending them. A few key pre-9/11 approaches include the following.

- GRAS (Generally Recognized As Safe). This designation is the result of the 1958 Food Additives Amendment. In practical terms, this amendment affected primarily food scientists responsible for recipe formulation, but also, to a lesser extent, the procurement function and manufacturing operations, and of course those manufacturing additives. Prior to 1958, the FDA did not formally have, nor was it authorized to address, the safety of additives. Most additives in use prior to 1958, which had not proven to be unsafe, were grandfathered in as GRAS. Since the amendment, the practical implementation of GRAS has been a collaborative effort between the FDA and the food manufacturing industry. The operative word in GRAS is *safe*, and in the context of GRAS, safe means that there is a reasonable certainty in the minds of competent scientists that the substance is not harmful under the intended conditions of use, and safety may be determined by scientific procedures or by general recognition of safety.

  In practice, only those responsible for food science, recipe formulation, regulatory compliance, and risk assessment are concerned with GRAS, and not those manufacturing and managing the manufacturing activities.

- GMP. Food safety for formulated food has been guided for many years by the U.S. FDA in the form of a set of guidelines called good manufacturing practices (GMP). Both food and nutritional supplements manufacturers shared common GMP until recently.

  GMP indirectly affects virtually everyone in the company not in a purely administrative function. It encompasses everything from plant layout to all manufacturing processes and ingredient and product storage. Think of GMP as specifications for repeatable, safe standard processes. These safe repeatable processes are a company's standard operating procedures (SOPs).

- SOP. GMP defined an outcome, not how to achieve it. How to satisfy the GMP was up to the company and detailed in the form of standard operating procedures (SOPs).

  Everyone involved in the manufacture of product, including factory and equipment maintenance, raw material receipt, inspection of raw materials, handling and storage of raw materials, product manufacture, and product inspection, is performing his or her work under the guidance of an SOP.

- Lot traceability. This has been with us for a long time, and it began with the agricultural produce industry in the late 1800s when refrigerated railcars were introduced, and permitted produce to be transported to distant population centers. Problems soon developed because of product perishability, and growers asked the U.S. Congress for legislation that started the evolution of food product traceability, but then the main purpose was to determine who was financially responsible for spoiled product, not food safety. In the 1990s the FDA, in response to incidents of food contamination, developed good manufacturing practices (GMP), in which one part of the guidelines focused on improving traceability, but this time with the central purpose being food safety. Subsequent experience brought to the surface additional food safety attributes requiring more precise information, and the information requirements expanded. Lot tracing does provide benefits beyond food safety: improved supply management, marketing, and reduced costs in the event of a safety recall.

On a day-to-day basis, lot traceability comes into play first with procurement (purchasing) that must contractually require a trace code for each incoming lot of raw materials from suppliers.

The next two departments that figuratively touch lot traceability are receiving and quality assurance, which must confirm that a lot trace code has been provided on all materials. Following that, those who store raw materials (the warehouse staff) need to keep lots segregated. When ingredients are issued to a production batch of an intermediate (aka subassembly) or final product, each must be recorded against a new, traceable lot code. After the batch has been produced, it is either stored in the warehouse again on a segregated basis or shipped to a customer. When shipping to a customer, the product's lot code/customer transaction must be linked and recorded. Finally, oversight for all of this is technically the responsibility of regulatory compliance/quality staff, but in reality everyone, starting with the CEO. A recap of those involved with lot traceability integrity includes the following:

1. Procurement (purchasing)
2. Incoming materials receiving
3. QA (quality assurance)
4. Warehousing (storage)
5. Warehousing (issue raw materials)
6. Production (record new lot code)
7. Warehousing (storage)
8. Shipping
9. Compliance/QA (oversight)

- HACCP. Hazard Analysis Critical Control Point is not a new system. The concept was developed in the 1960s by the Pillsbury Company, while working with NASA and the U.S. Army Laboratories to provide safe food for space expeditions.

  HACCP is a preventative food safety system that through the careful analysis and understanding of the behavior of food ingredients throughout processing, handling, and storage is designed to ensure safe food products are delivered into the food supply chain. It is not a traditional quality control inspection system that features inspection for defects or prevention of defects, although in practice both approaches are often used. After its development in the 1960s, HACCP migrated in the late 1970s to specific food industry sectors in which its use became mandatory, and it is widely used today in other sectors in which its deployment is optional, often for marketing purposes, to differentiate a product from those of its competitors.

Because the underlying logic of HACCP is sound, widely implemented, and knowledgeable expert resources are available to help develop and implement a plan, it has become internationally recognized as the best system for ensuring food safety and is endorsed by many regulatory authorities outside of the United States. The ISO 22000 standard merges HACCP and ISO 9000 (quality system).

A chronological history of significant pre-9/11 food safety events follows:

- 1200: King of England establishes first food law.
- 1862: USDA established.
- 1880: The USDA establishes the cornerstone in what eventually evolves into lot traceability.
- 1906–1907: The original Food and Drugs Act, the Meat Inspection Act, is passed by the U.S. Congress and First Certified Color Regulations.
- 1930: The McNary-Mapes Amendment authorizes FDA standards of quality for canned food, excluding meat and milk products, and the name of the Food, Drug, and Insecticide Administration is shortened to the Food and Drug Administration (FDA).
- 1933–1939: FDA recommends a complete revision of the obsolete 1906 Food and Drugs Act, and in 1938, the Federal Food, Drug, and Cosmetic (FDC) Act is passed by Congress, which required safe tolerances to be set for unavoidable poisonous substances; standards of identity, quality, and fill of container for foods; and authorized factory inspections. In 1939, first foods standards were issued.
- 1940: FDA separated from the USDA.
- 1943: The U.S. Supreme Court ruled that the responsible officials of a corporation may be prosecuted for violations and it need not be proven that the officials intended, or even knew of, the violations.
- 1949: FDA publishes guidance to industry, "Procedures for the Appraisal of the Toxicity of Chemicals in Food."
- 1950–1960: Congressional investigation of the safety of chemicals in foods, laying the foundation for the 1954 Miller Pesticide Amendment, the 1958 Food Additives Amendment, the 1960 Color Additive Amendment, and the Factory Inspection Amendment.

- 1958: Food Additives Amendment enacted, requiring manufacturers of new food additives to establish safety. The FDA publishes the first list of substances generally recognized as safe (GRAS) along with Animal Drug Amendments.
- 1969: The FDA begins administering sanitation programs for milk, shellfish, and food service, known as HACCP and interstate travel facilities.
- 1970: The Environmental Protection Agency (EPA) was established. It took over the FDA program for setting pesticide tolerances.
- 1982: The FDA publishes *Red Book: Toxicological Principles for the Safety Assessment of Direct Food Additives and Color Additives Used in Food.*
- 1984: Fines Enhancement Laws of 1984 and 1987 amend the U.S. Code to greatly increase penalties for all federal offenses for individuals and corporations. Note: See 1943.
- 1994: Dietary Supplement Health and Education Act defines dietary supplements and dietary ingredients as food (Figure 3.2).

Food safety laws and practices in countries outside of the United States varied greatly prior to 9/11. The United States historically has been a multicultural nation and is a federation making the formulation and

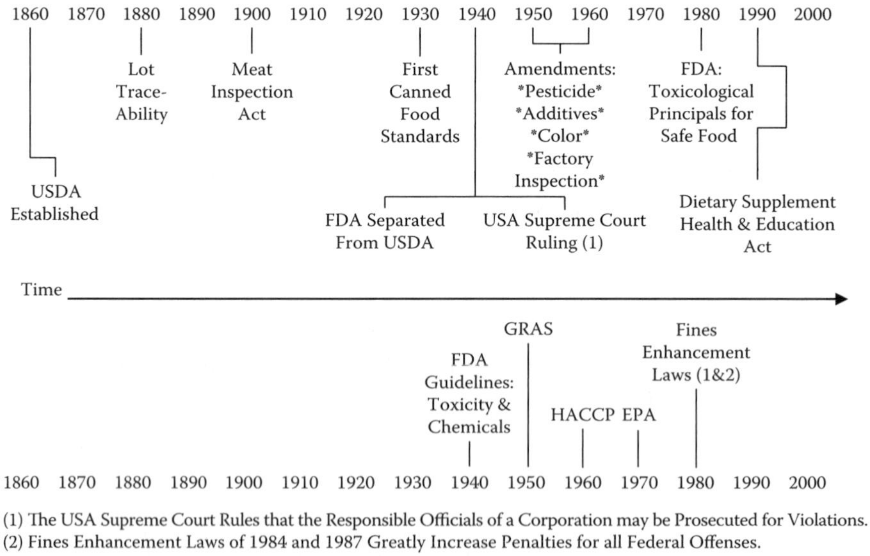

FIGURE 3.2
Key food safety developments in the United States: 1800–2000 timeline.

implementation of common food safety laws easier than other nations, but in some cases more difficult when compared to cultures with a specific defining culture, such as the culinary preferences of a German versus a Greek. For two other examples:

- Europe has historically been a collection of different nations with different cultures and views on food. Large areas of Europe had previously been unified by empires built by force. The march toward a unified Europe that enhanced trading and economics but preserved cultural views started in 1946 with the first major collaborative arrangement to implement economic integration in the form of the European Economic Community (EEC). The six founding country members were France, Western Germany, Italy, Netherlands, Belgium, and Luxembourg. Because of its success, the EEC attracted new member nations, integrated more shared laws and standards, and evolved into the European Union (EU). Until 2002 most of food safety law (this is a generalization) was entrusted to each sovereign government. Since the EU has more than twenty member countries, plotting their histories prior to 9/11 would be information overload.
- Japan has a very distinct culture and views of the purpose of food with strong emphasis on health and well-being.

It is also an island nation that doesn't have domestic food self-sufficiency and relies on food imports. One estimate is that it can only produce, on a caloric scale, forty percent of its needs.

THE BIOTERRORISM ACT: A NONTECHNICAL OVERVIEW

The 9/11 terrorist attack on the World Trade Center was the catalyst for the Bioterrorism Act in the United States, which is best described as a top-level set of goals that the U.S. Congress directed the Secretary of Health and Welfare to enforce at a detailed level with agencies such as the FDA. Although the Bioterrorism Act itself is quite broad, the focus of our conversation is on the discipline needed to adhere to the record-keeping requirement (records maintenance) in support of lot traceability, which given the potential gravity of a tainted food recall, is significant. Figure 3.3 is a simplified view of Title 111 of the act, which highlights the

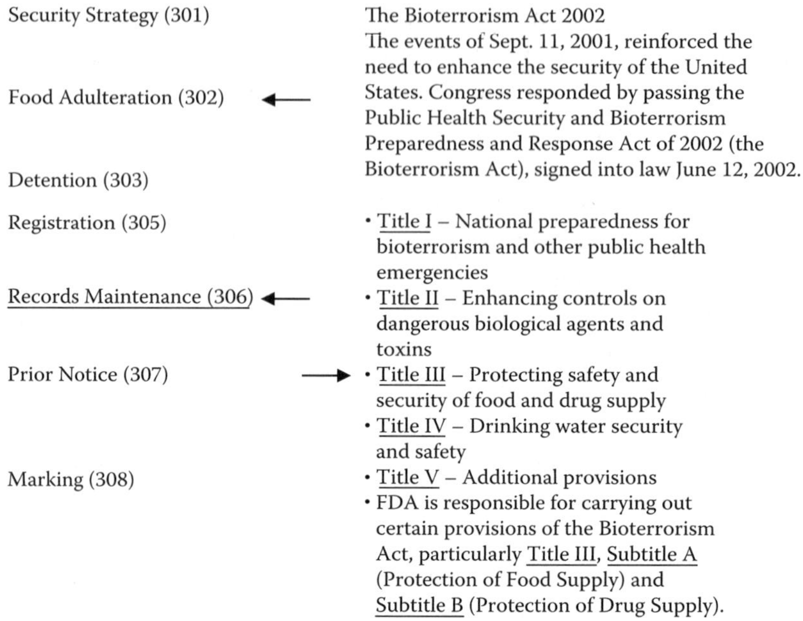

FIGURE 3.3
USA Bioterrorism Act overview.

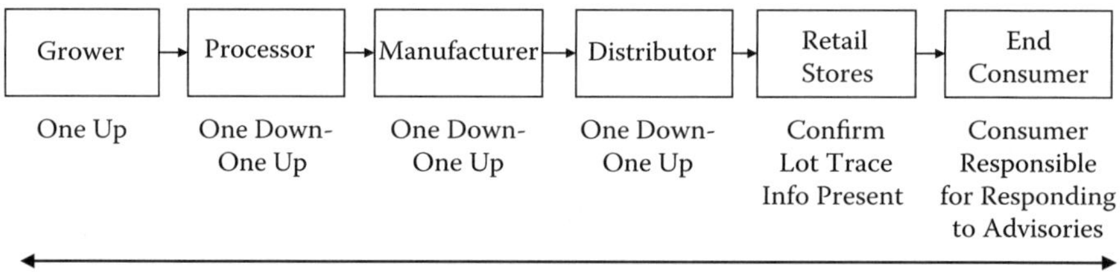

FIGURE 3.4
Lot tracking across the food supply chain. One-up and one-down accurate record keeping.

responsibility for lot traceability across the supply chain. Exactly where accurate lot traceability is required across the supply chain and internally at a food processor/manufacturer is shown in Figures 3.4 and 3.5.

FOOD SAFETY OUTSIDE THE UNITED STATES

Food safety thinking outside the United States has also been influenced by the 9/11 terrorist attack, and in the case of the European Union's progress

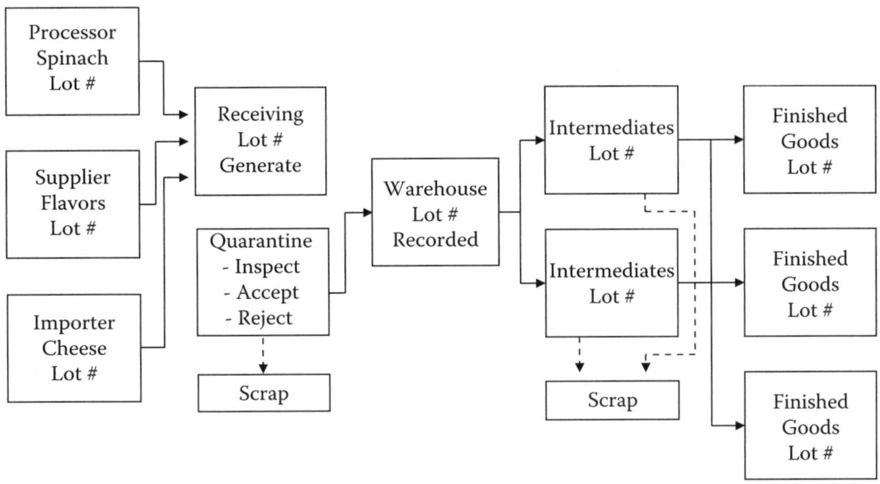

FIGURE 3.5
Example: Internal lot trace.

toward common laws, resulted in the development of the European Union Common Food Law.

A partial list of the principles and objectives in the law follows:

- Ensures a high level of protection of human life and health
- Establishes the rights of consumers to safe food and accurate information
- Creates a framework to ensure a coherent approach in food legislation
- Harmonizes existing national requirements to ensure the free movement of food and feed in the EU
- Takes international standards into consideration, except where this might undermine a high level of consumer protection
- Establishes common definitions, principles, and obligations covering all stages of food/feed production
- Establishes the principles of risk analysis in relation to food
- Establishes the structures and mechanisms for the scientific analysis undertaken by the European Food Safety Authority (EFSA)
- Establishes that the three interrelated components of risk analysis (risk assessment, risk management, and risk communication) provide the foundation for food law
- Provides full transparency of legislation and the evaluation and explanation of potential risks with effective public consultation about food safety

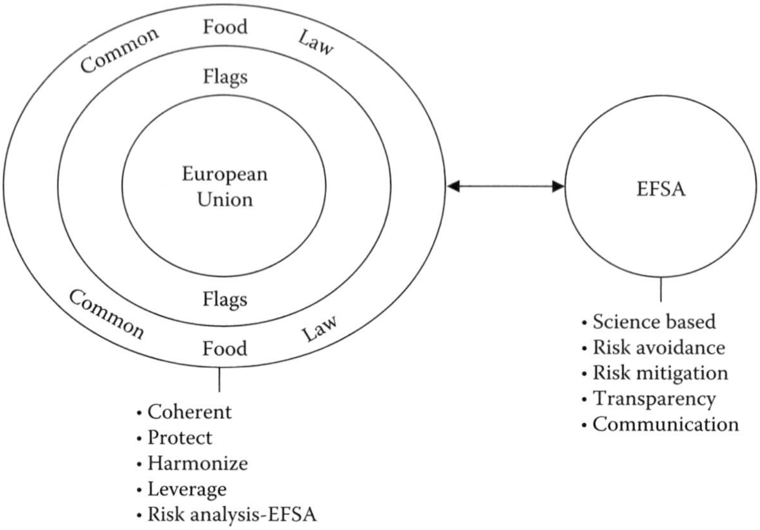

FIGURE 3.6
EU common food law and EFSA.

The law establishes the principle of using science and independent scientific advice that led to the creation of the European Food Safety Authority (EFSA). EFSA says the following on its website regarding its charter and responsibilities:

> The European Food Safety Authority (EFSA) is the keystone of European Union (EU) risk assessment regarding food and feed safety. In close collaboration with national authorities and in open consultation with its stakeholders, EFSA provides independent scientific advice and clear communication on existing and emerging risks.

Figure 3.6 depicts the intent, roles, and synergy of the law.

GOING BEYOND THE BIOTERRORISM ACT: THE BIOTOXIN ATTACK SIMULATION

After 9/11, key food industry thought leaders in the United States became concerned about the human health consequences of a possible, intentional contamination of food with a bioterrorism agent at some vulnerable point in the distribution chain. This concern was transformed into a strategy of developing simulation tools using scientific data and computing technology to help all the stakeholders in the food supply chain simulate a biotoxin attack, understand its behaviors, and put in place, where possible, preventive measures. Also, if a bioterrorism attack did occur and the agent used was identified, the strategy explained how to plan, manage, track, contain, and start remediation activities. The envisioned stakeholders included not only food producers, but also local, state, and federal government agencies, healthcare providers, public health departments, and academia.

The vision and strategies had great appeal to the FDA, which had started to think in terms of analysis tools to give to the food manufacturing community, which simulated attacks on our food supply chains using various contaminating agents. The goal was to heighten food industry awareness and motivate all members of a supply chain to put in place preventive measures and recovery plans in the event of an attack. At this point, collaboration started between the food industry and various U.S. government agencies, which created a shared vision and initiative. The expression "food defense" entered into the industry vocabulary and supplemented existing food safety thinking.

Two easy-to-use software systems were developed. One is the Consequence Management System, and it visually depicts the evolution and potential outcomes of food contamination events. With this system, authorities and food companies could determine what monitoring systems needed to be put in place and what policies were needed. They could develop appropriate plans and determine what education was required to establish a state of readiness in the event of an attack with specific toxic agent. In conjunction, an integrated database was developed using actual data provided by the food industry, industry experts, medical and public health professionals, and government agencies—FDA, Department of Homeland Security (DHS), USDA, and Centers for Disease Control and Prevention (CDC). This database was needed because up to that point, testing wasn't routinely done for these toxic agents, nor were their symptoms and life cycle stages clearly understood by all the food safety stakeholders. The development of this database clearly highlights the power of collaboration.

A companion system was developed that focuses on responding to an attack and supports four areas:

- Managing and tracking
- Containment

- Remediation activities
- Sanitation and cleaning protocols

Sometimes a tool proves useful for purposes beyond its initial design. This is the case for these simulation tools and their companion database. They can help those in a food supply chain avoid food safety crises unrelated to intentional terrorist attacks. The availability of these tools was widely broadcast to the food industry and other supply chain stakeholders in the mid-2000 timeframe.

RESULTS AND DEVELOPMENTS SINCE THE BIOTERRORISM ACT

A few significant developments transpired since the passage of the Bioterrorism Act:

- Increased importation of food
- Establishment of the European Union Common Food Law
- Increased food product recalls
- Highly publicized food safety episodes
- Development of separate good manufacturing practices (GMP) for food supplements
- U.S. Food Safety Modernization Act (S.510)

INCREASED IMPORTATION OF FOOD

Importation of food in both the United States and EU has increased. Japan has in modern times been a major importer of food. Numbers are often misleading. Is the increase based on volume or value? This increase has been driven by the acceptance that:

- Global food trade is OK.
- Consumer preferences for foods out of season locally need to be satisfied.
- Consumer demands for food variety, including ethnic foods, need to be satisfied.

Another contributor is the low cost of logistics/transportation—imports/exports are exempt from fuel tax by international agreement. Technology is another contributor; we can now extend the freshness of certain foods, mitigating the natural time-distance limitation. Profit is also "fuel" for importing food. If a profit is there, someone will go get it.

In any import discussion, the People's Republic of China (PRC) invariably comes up because of the food safety incidents experienced in markets they export to. During the 2000 decade, food imported from the PRC to the United States increased three hundred percent. This translated into one percent of the total food consumed, but based on the public's perception, it was much more. Most of this was fish and shellfish. The PRC is a large, complex country with more than two hundred million farms and food processors. Because of this, anyone dealing with a supplier from the PRC needs to be alert to the adherence to food safety concepts, methodologies, and techniques by his or her supplier.

THE ESTABLISHMENT OF THE EU COMMON FOOD LAW

The EU Common Food Law was discussed earlier. It's reasonable to conclude that the U.S. 9/11 incident contributed to this law. But, the EU was on an evolutionary path that might have generated the law on its own, independent of the 9/11 terrorist attack. The EU has some advantages in formulating food safety laws; it has to reconcile different cultural views of food and food safety. There are only two possibilities: Those with a more stringent standard elevate the standard for all, or those with a lower standard reduce it for all.

INCREASED FOOD PRODUCT RECALLS

The purpose of the record-keeping, lot traceability requirement is to facilitate rapid, precise, and efficient recalls of tainted food. Historically the FDA has collaborated with food manufacturers to initiate recalls. Although a number of years have passed since the mandated implementation of this record-keeping requirement, there is significant opportunity for improvement. As an example, one government audit of the completeness of lot

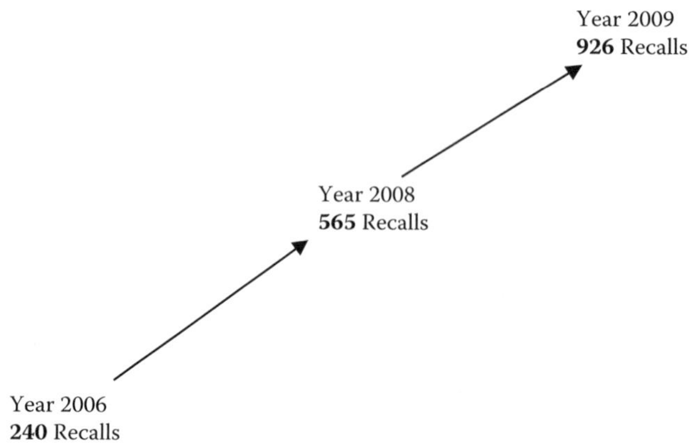

FIGURE 3.7
Increased food recalls since the Bioterrorism Act.

trace records of forty food products found that only five had full trace records. Also, since the passage of the Bioterrorism Act, the number of recalls has increased dramatically (Figure 3.7).

Aside from the moral obligation to protect the food supply chain, there are significant economic consequences for inaccurate lot trace records. One of the major themes in this book is that the accuracy of those records and rapid access to them in an emergency is gravely important, and also how this can be leveraged into superior inventory and supply-chain management, which will be covered in subsequent discussions in this book.

HIGHLY PUBLICIZED FOOD SAFETY EPISODES

There have been some dramatic food safety episodes since the U.S. Bioterrorism Act. Normally, we expect animal meat products to be the most problematic, but spinach, tomatoes, pet food, peanut butter, pizzas, and eggs—among other non-animal meat food products—became the center of attention on news outlets and received extensive media coverage in both print and telecast form.

In some cases, the economic consequences have been devastating. Some industry experts, using the recall in 2006 in North America for *E. coli* contamination as an example, believe it can take years for a micro-industry to fully recover. In some cases, major food suppliers have been shut down and gone out of business.

Because we now live in a world that has many information sources pushing information out electronically in real time, the consumer has become sensitive to food safety issues and demands solutions to these issues. The number of highly publicized recalls and the increase in the overall number of recalls probably were the catalysts for the U.S. Food Safety Modernization Act (S.510).

Another contributor to food safety episodes is the globalization of the food supply chain. Because we desire the importation of safe food, and each country has differing views of food safety and how to handle the balance between safety and corporate profitability, managing locally is easier than managing nationally, and managing nationally is easier than managing globally.

In summary, food safety episodes will be with us for a long time. They can be avoided by making food safety important to every worker and manager in every company in the food supply chain. Having every worker and manager contribute and enabling them via education and leadership are key. Ignoring this has no corporate or personal benefit, only negative consequences.

DEVELOPMENT OF SEPARATE GOOD MANUFACTURING PRACTICES (GMP) FOR NUTRITIONAL SUPPLEMENTS

Nutritional food supplements have an interesting history. Until the 1960s they didn't exist in the general public's view. Food supplements, also now known as nutraceuticals, were very basic, and for an iconoclastic consumer were the answer to the industrialization of the food supply. Putting things in context, during that period the pesticide DDT was considered safe and was widely used. Ignorance is not bliss. The then typical "super" supplements were the B complex and C complex vitamins and various oils, both vegetable and fish based. As an iconoclastic health-conscious consumer (then and continuing), the author remembers those days well.

For a long period of time food supplements existed in a figurative no-man's land. Exactly what was it? Eventually the U.S. FDA defined supplements as food and the food GMP applied to their manufacture.

Since then the number of nutritional supplement products has increased tremendously, and many of the formulations have multiple active ingredients. Some of them are based on traditional preparations from India and China, both of which have a rich history of use for health (preventive medicine) and treatment of illness. In fact, many of the active ingredients in any supplement are sourced from outside the United States. Recent surveys show food supplements have been generally accepted by the public, with approximately fifty percent of the U.S. population using them.

Unfortunately, a small number of supplement manufacturers made unsubstantiated health claims, and independent laboratory testing often found that the actual potency found in testing was less than the declared potency. In the case of a botanical ingredient, it is difficult to get a consistent potency, but the disciplined and quality-conscious manufacturer puts controls in place to manage this variability.

With the growing popularity of nutritional supplements, the U.S. Congress put into law the Dietary Supplement Health and Education Act (DSHEA). The FDA conducted studies to determine if the food GMP were adequate for food supplements. FDA concluded they weren't and developed specific GMP for food supplements. These new GMP have very tight controls embedded in them, and the FDA developed approximately a two hundred-page preamble to the actual GMP to help the manufacturer understand its obligations. Many quality management professionals were impressed with the effort the FDA made in launching the new GMP.

Despite operating under two separate GMPs, there is an increasing crossover of nutritional supplements into food formulations because food manufacturers increasingly offer "functional foods." A functional food is any healthy food claimed to have a health-promoting or disease-preventing property beyond the basic function of supplying nutrients. An ancestor of functional food is milk with the addition of vitamins A and D. The number of food products that meet the functional food definition is rapidly increasing, and the definition helps processors differentiate traditional products in a competitive environment, vying for the health-conscious consumer.

There is a slow but steady trend today for the more demanding guidelines to migrate to industries that currently have less demanding guidelines and regulatory obligations.

U.S. FOOD SAFETY MODERNIZATION ACT (S.510)

There is an old adage: "Common sense is an uncommon quality." This act finally puts a dose of common sense into food safety in the United States. Prior to it, the FDA couldn't initiate a recall, but now it can. It also requires a safety analysis by food producers. Consider it half of a HACCP plan. More site inspections are mandated. But it is only a milestone on the journey to a dependable safe food supply chain. Expect more in the future. Only companies and their managers who aspire to be class A in conformance in food safety will survive the more disciplined food safety regulatory future.

SUMMARY

This chapter is not directed to the regulatory compliance or quality assurance professional but those who work with them. Compliance and quality professionals should be able to build on this overview and fill in the gaps with their coworkers. Thank you for reading this chapter and being a guardian of food safety. Keep it up! We all appreciate your concern, efforts, and professionalism.

QUIZ

3.1. Food supply chains have been increasingly localized.
☐ True ☐ False

3.2. The challenges in ensuring a safe global food supply chain are minimal.
☐ True ☐ False

3.3. In the United States the responsibility for ensuring food safety rests exclusively with the FDA.
☐ True ☐ False

3.4. Prior to the Bioterrorism Act there were no meaningful food safety laws.
☐ True ☐ False

3.5. Lot traceability is a recent concept.
☐ True ☐ False

3.6. GMP outline a set of requirements and SOPs are how an organization will satisfy them.
☐ True ☐ False

3.7. HACCP is designed around the concept of inspecting for defects.
☐ True ☐ False

3.8. The U.S. Bioterrorism Act doesn't address lot traceability.
☐ True ☐ False

3.9. There are no analytical tools to help prepare for an intentional attack on a food supply chain.
☐ True ☐ False

3.10. The United States has been the only country concerned with food safety.
☐ True ☐ False

3.11. Food and nutritional supplements manufacturers operate under one common GMP.
☐ True ☐ False

3.12. Food safety requirements continue to evolve.
☐ True ☐ False

The answers to these questions can be found in the Appendix along with definitions, website addresses, and recommended reading.

4

The Top Ten Issues in Food Manufacturing

This chapter examines an annual survey from *Food Engineering*, a major and respected food publication dedicated to food manufacturing that identifies the industries' top ten operating issues. In this survey, the number one issue was food safety. Business issues within an industry tend to remain semi-static unless there is a disruptive event, such as:

- New, demanding, government-imposed regulation
- A serious economic decline
- New technology
- Dramatic changes in global trading
- A significant (hundred-year) natural disaster
- The impacts of global warming

Absent any disruptive event in any given year, the issues will essentially remain the same from year to year.

All ten issues from this survey will be examined in detail. In the published survey results, each issue was treated by the respondents as a stand-alone one. An examination of the issues clearly indicates that satisfying eight of the issues was a prerequisite to managing the larger issue of food safety. Netting it out, nine of the top ten issues were interconnected. Seeing the connections between the various issues required an understanding of cause-and-effect relationships, which is a different thought process than is typical in many organizations today.

This chapter illustrates the connectivity between the issues and puts together an appropriate strategy and creates a priority ranking for dealing with the issues, by answering the question: Where does one start the improvement process, and what is the appropriate sequence of improvement projects?

More important is the need to think differently. What is the context? Are we addressing a symptom or a root cause? Are there enabling processes required to mitigate the issue? This chapter has many thought-facilitating graphics because a picture is worth a thousand words.

The top ten issues are:

1. Food safety
2. Automation
3. Skilled labor
4. Product (lot) traceability
5. Systems integration
6. Maintenance improvements
7. Supply chain integration
8. Increased capacity
9. Employee training
10. Improved changeover

Since food safety is the number one priority, which of the other nine most directly impact achieving it? If we confine the analysis to the top ten issues, there are five (Figure 4.1):

- Product traceability (issue 4)
- Skilled labor (issue 3)
- Maintenance improvements (issue 6)
- Employee training (issue 9)
- Automation (issue 2)

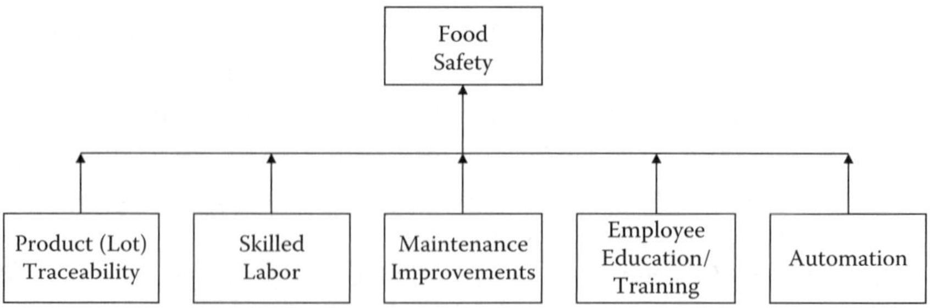

FIGURE 4.1
The five issues that impact food safety.

PRODUCT (LOT) TRACEABILITY

Lot traceability has been with us in the United States in one form or the other since the 1880s (see Chapter 3). It's been an assumed requirement for either business liability or food safety purposes for many decades. In some cases it has been an informal business accommodation between trading partners. In today's marketplace its purpose is to identify what product unintentionally or intentionally (bioterrorism) must be rapidly recalled, that is, withdrawn from the food supply chain because it's been corrupted and is dangerous to the consumer. Unfortunately achieving lot traceability has been elusive. Since the passage of the USA Bioterrorism Act under the record-keeping requirement, it has been mandatory. Despite it being a legal requirement, it still has proven to be challenging based on audits for full traceability. The record-keeping requirement in the Bioterrorism Act only requires one-up, one-down traceability record keeping. Seems simple:

- Record all the lots of incoming raw materials and their required documentation.
- Record what raw material lots were used to produce a final product.
- Record what customer was shipped the final product.

A simplified illustration is shown in Figure 4.2.

In practice, keeping track can be difficult because:

- A raw material lot can (and often is) used on multiple batches of intermediate (think subassembly) *and* final products.
- Intermediates are often consumed by multiple end-item batches.

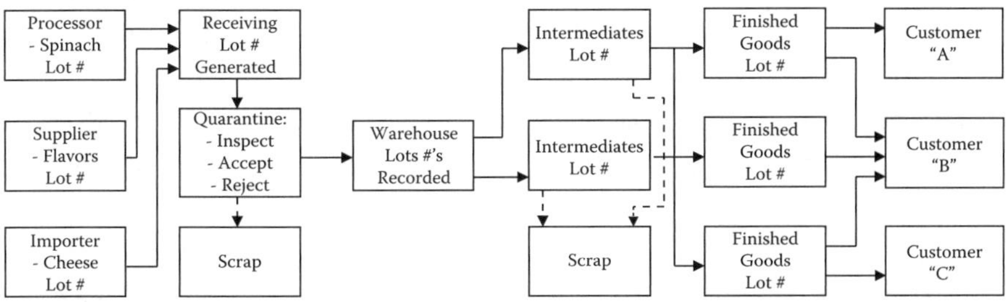

FIGURE 4.2
A lot tracing example.

- End-item products are shipped to more than one customer from a production end-item batch.
- Incoming raw materials aren't really from a linear supply chain but from an intricate supply web.

The difficulty in achieving accurate and complete lot traceability is underscored by a sample audit by the U.S. Inspector General, in which the majority of traced lots failed to provide a complete history. Without accurate and complete lot traceability, it isn't possible to conduct an effective and focused product recall of dangerous product lots. A by-product of this lack of data integrity is the economic consequences of having to recall more product lots than truly necessary. But most important is the danger to the consuming public. Without accurate and complete lot traceability, a company cannot satisfy the obligation to provide safe food.

But why is achieving accurate and complete lot traceability so difficult? Part of the reason is the complexion or makeup of the food and nutritional supplement industry itself. Although there are many large companies, the industry is populated with tens of thousands of small- to medium-sized companies, many of which are highly specialized. Also, there are a high percentage of contract manufacturers.

It is not unusual for these smaller firms to use a manual, paper-based approach to record keeping or to use computerized spreadsheets. Often the planning and control in the smaller firm is a mix of paper-based systems, numerous spreadsheets, and computerized financial ledgers. There are four problems with this approach:

- There often is a time lag between the activities (transactions) in receiving, QA, manufacturing, and shipping. The experience of other industries outside of food and nutritional supplements has shown that a delay in posting a transaction can result in inaccurate data for a variety of reasons.
- Since there are multiple databases there is no easy way of identifying conflicting data, i.e., that lot number was already used on a different batch and product.
- There is a separation of the person generating the data and those recording it for record-keeping purposes. Not only does this have the potential for inadvertent inaccurate data, but also accountability for data accuracy. Another downside is additional administrative costs.

- With these disparate data sources, paper and spreadsheets, can the company pull a complete lot trace history when demanded to do so by the U.S. FDA or important customers, e.g., Wal-Mart, in the required time?

OK, what about large companies? They have had the advantage of being able to invest in process automation technology, information technology (computing systems), and employee training. This doesn't mean they all perform at a world-class level of excellence. Maintaining performance requires persistence, vigilance, and a commitment to continuous improvement. The gap between large and small companies has narrowed relative to regulatory compliance, or at least there is the opportunity to do so. How so?

Information technology, which historically has been very expensive, is now relatively inexpensive. Over the last decade, the cost of computers, computer networks, planning and control and record-keeping application software, bar code scanning software and equipment, and maintaining all of it has dramatically decreased. This is the result of the maturing of the information technology industry and the demands of customers who really want a service without all the complications previously associated with using information technology. There is another wave of cost reduction and simplification in using technology on the way. Organizations now want what once required a substantial investment to be a service, much like the consumption of electricity.

In terms of historical perspective, large companies functioned as a research and development (R&D) lab, and once the investment sorted out what worked from what didn't, and that segment (large companies) absorbed everything, a trickle-down effect started and those who provided these information technology innovations began to recalibrate their products and services to serve the smaller organization.

Here are a few examples of the trickle-down effect and commoditizing of information technology and equipment:

- Relational database management systems (RDBMS) are now cost-effective even for the individual consumer. This allows the linking of the trace history and real-time editing for data errors; i.e., this lot number has not been released by QA.
- Enterprise Resource Planning (ERP), a complete and fully integrated planning, reporting, controlling, and centralized data repository software system, has been developed for the smaller company that is required

to maintain lot traceability and has become cost-effective to own and use and historically has produced an excellent return on investment (ROI). The good news is that there are numerous software developers servicing the food and nutritional supplements manufacturer, which means a company has choices. Many of these ERP systems sit on top of a RDBMS. Another benefit is that only authorized workers and staff have access to the ERP system since it has security features built in.

- Bar coding and data collection equipment has substantially come down in price, and with increased durability, allowing for instant data updating at the point where work is performed with no added administrative cost. Every time someone enters some data via a keyboard or on paper there is the potential for making a mistake.
- We now live in a wireless world, and the costs of most electronic devices enabling connectivity have decreased. This means a worker doesn't have to go to a personal computer (PC) or terminal to input data.
- Software as a service (SaaS) is becoming more commonplace. The benefit is that the customer, for example, using an ERP system, doesn't have to own a computer server, hire information technology staff, maintain the ERP system, etc. All that is needed is a simple computing device and access to the Internet.
- The World Wide Web (WWW) has the potential for cost-effective delivery of education and training content to a company's workforce on best practices, specifically the various techniques used to achieve and maintain data accuracy and integrity, the most important enabler to effective lot traceability.

  One approach to using the WWW for educational purposes is the use of webinar technology. An instructor could be in any part of the world while the student is in another. Most webinar systems can record the "live" initial session and be replayed at the convenience of the student. Webinar systems are offered on a SaaS basis and are inexpensive.

  For the instruction of more complex techniques, webinar technology is combined with a learning system to educate and test the student's learning without the need for a set time and date. The system keeps track of progress and the student gets the benefit of flexibility. The benefit of this form of educating is that travel costs are avoided.

The net of all of this is that accurate and complete lot traceability is essential to food safety, and the tools to achieving it are available to even the small manufacturer.

SKILLED LABOR

All high-performance organizations invest in their workers at all levels, starting with the CEO and continuing throughout the organization. Every worker, no matter his role, has an obligation to protect the safety of the food supply chain. It doesn't require a major mistake to corrupt food and make it unsafe. The skilled worker is an important ingredient in achieving food safety.

Unfortunately, for a manufacturer there will always be a shortage of available-to-hire, skilled workers. This is because of the size and complexity of the food and supplements industry. On the surface, making food looks simple, but a lot of complex science is below the surface. This means that the typical manufacturer has to develop its own skilled workforce or risk possible consequences.

But how?

1. Recognize that a skilled (and semiskilled) workforce can be developed.
2. There is no other option but to commit to "growing your own."
3. Commit to having one or more (a size of company consideration) individuals dedicated and accountable for developing a skilled workforce. This individual or team needs to have the visible and continuing support of the CEO. Is human resources (HR) the right choice for this responsibility?
4. Profile your workforce. Who has English language proficiency? Who can read (any language)? Who has a good work ethic?
5. Invest in your skilled labor leader. Have him or her educated in the adult learning principles. The American Society for Training and Development (ASTD) is a good resource.
6. If your profile shows that there are individuals that can't read, develop education and training materials that are picture based and supplemented by verbal instruction.
7. If your workforce has a significant number of workers who don't have English language skills, translate the materials into their language.
8. Revisit your sanitation standard operating procedures (SSOP) and make sure they are in all appropriate languages, are clearly and concisely written, and have illustrations. Do a walkthrough—there will be surprises.

9. Work with each worker and create a career path, including the education and training required and time lines. This must be done at least annually if not more often. When this is done an informal contract is established, expectations agreed, and a sense of mutual commitment shared.
10. Create mini-tests to confirm knowledge and conduct them at regular intervals. Continue to invest in the trusted worker.

The figurative "ten commandments" of developing and maintaining a skilled workforce is not rocket science or expensive, and the ROI is attractive.

MAINTENANCE IMPROVEMENT

One issue in the survey was maintenance improvement for the purpose of keeping machinery running to reduce downtime and reduce maintenance costs. Poorly maintained equipment can also be a contributor to producing unsafe food. In some companies equipment is "maintained" only when the machine or device breaks down. The downside of this approach from a food safety perspective is the product recall liability. There have been many product recalls for such reasons as metal in the food product from grinders whose internal parts started to deteriorate, small glass fragments because of malfunctioning filling devices, spoiled product because it wasn't processed at the correct temperature because the equipment's controls weren't accurate, and the list goes on.

For companies operating under Hazard Analysis Critical Control Point (HACCP) the risk associated with the potential of malfunctioning equipment is required, as is a risk mitigation plan. Even for those currently not obligated to do a full HACCP, there is the possibility that it will be required for all manufacturers and distributors in the future.

The good news is all three needs, food safety, equipment uptime, and containing maintenance costs, can be satisfied. There are four approaches available:

- Implement 6S.
- Frequent assessment of operating conditions by the machine or device operator.

- Implement a formal preventative maintenance program based on engineering input from the machine manufacturer or, in the case of custom-designed equipment, internal engineering staff.
- Adopt the single-minute exchange of die (SMED) thought process.

Implement 6S

6S has its origins in the approach to Lean manufacturing made famous by Toyota. Until recently it was called 5S. 6S is very complementary to cGMP and SSOPs. The objective of 6S is simple: Take a disorganized work area and make it organized, clean, visual (it is obvious as to what is done here), and safe. 6S is:

1. **Sorting:** Eliminate all unnecessary tools, parts, and instructions. Keep only essential items and prioritize them and keep them in an easily accessible place.
2. **Straightening or setting in order, stabilize—simplify:** Make a place for everything and everything should be in its place. The place for each item should be clearly labeled. Items should be arranged in a manner that promotes efficient work flow.
3. **Systematic cleaning:** Keep the workplace tidy and organized. Maintaining cleanliness is a part of the daily work—not an occasional activity initiated when things get too messy.
4. **Standardizing:** Work practices should be consistent and standardized. All workstations for a particular job should be identical. Everyone should know exactly what his or her responsibilities are for adhering to the first 3Ss.
5. **Sustaining the discipline or self-discipline:** Maintain and review standards. Once the previous 4Ss have been established, they become the new way to operate. Maintain focus on this new way.
6. **Safety:** Is the primary concern not only for the food being made but also for the workers producing it?

6S isn't just for manufacturing; it is a philosophy appropriate for every department in the enterprise.

Frequent Assessment of Operating Conditions by the Machine or Device Operator

A few decades ago the manufacturing community "discovered" total quality management (TQM). It was a multifaceted approach to quality. One of the tenets of TQM was that a worker could be trained to do routine equipment cleanup, equipment operating assessment, and minor maintenance. Of course all of this with forethought from engineers and a simplified checklist approach. Well, history has shown that the TQM advocates were right and the operator can keep the machine running effectively and make a contribution to food safety.

Implement a Formal Preventative Maintenance Program

This sounds logical since unscheduled equipment downtime is expensive and disruptive to many organizational processes, such as capacity planning, scheduling of work, and worker assignments, and if the order can't be shipped on time, customer service has to (or should) notify the customer. A preventive maintenance program is a key contributor to an effective TQM program. Given the impact of unplanned equipment downtime, one would think every company would have a well-thought-out, engineering-based, concisely described maintenance plan strictly adhered to for each piece of equipment. But, common sense is an uncommon quality, and many organizations have a preventive maintenance program that has lots of room for improvement.

Adopt the Single-Minute Exchange of Die (SMED) Thought Process

SMED was developed not to necessarily improve the maintenance of equipment but to reduce equipment setup times, which reduced the economic lot size of a batch, and hence reduced inventory investment because a small lot size could be economically produced. Inventory is not cheap, and carrying excessive inventory brings with it under the best of circumstances an additional (carrying) cost of twenty-eight percent. The value of SMED to improving maintenance is its core logic; think through the changeover process, try to simplify it, organize everything needed to perform it, and perform it the same way each time. Less time and cost amplifies the benefit of preventive maintenance.

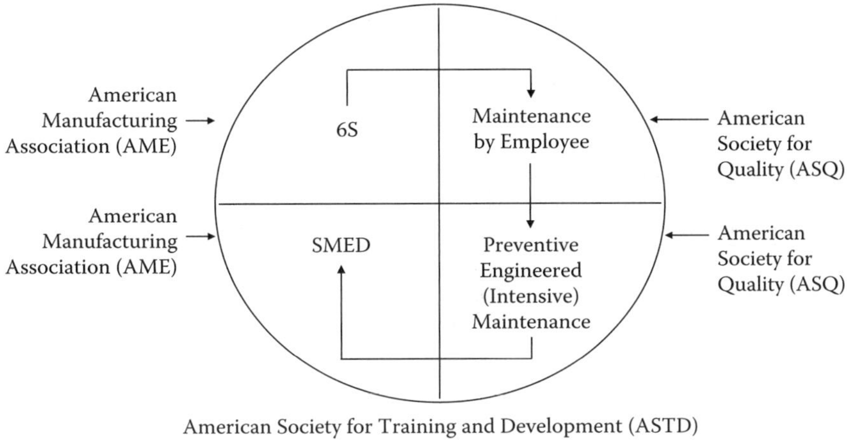

FIGURE 4.3
Sources for external education and guidance, and internally developed courseware.

The principles for creating cost-effectiveness by improving equipment maintenance have been codified into a body of knowledge based on the experiences of thousands of companies. Also, there are cost-effective ways of learning about the proven underlying principles and how to implement best practices. There are four concepts and "how to" resources that are available to any company: universities, professional associations, industry associations, and private, for-profit educators. Professional associations tend to add and extend the body of knowledge and are often nonprofit and generally less expensive. The important point is that bringing into an organization the necessary skills to be competitive *and* manufacture safe food isn't difficult or expensive; it's a matter of executive leadership and discipline (Figure 4.3).

EMPLOYEE TRAINING

Employee training to grow a skilled labor force was discussed earlier, but there is also a need to train semiskilled workers who are the majority of employees. There is a saying: "A chain is only as strong as its weakest link." The difference between the needs of the skilled worker and those of the semiskilled worker is in the depth of education and training. What is needed is that work be carefully designed and performed the same way every time. Practitioners committed to Lean thinking (aka JIT) call this

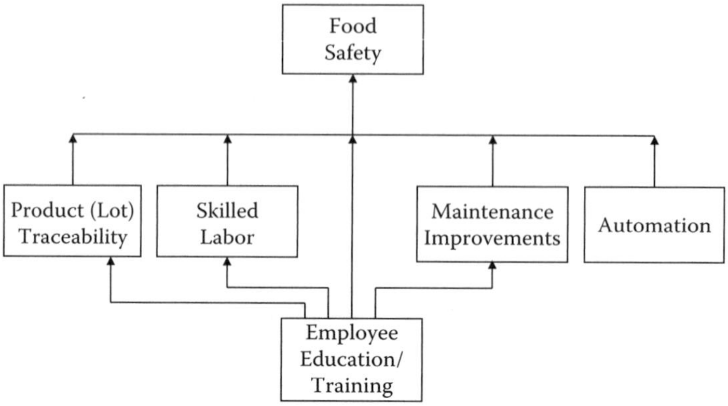

FIGURE 4.4
Education and training is required for all employees.

standard work. SSOP is a form of standard work. For those committed to HACCP, it also exists in the form of a critical control point (CCP) in which work procedures are specifically designed to maintain food safety. What are often missing are work steps that aren't SSOPs or CCPs. These can't just be verbal instruction from a supervisor or departmental lead. Standard work must be designed, carefully and skillfully documented, workers trained on it, and the steps rigorously adhered to. It gives the supervisor the framework needed to be an effective leader. An important element in training is to make sure everyone knows why it needs to be done that way and the consequences for failing to do so in terms of food safety or financial impact to the company.

Earlier a figurative ten commandments of developing an organizational capability to educate and train and grow an effective workforce was discussed. Nothing has changed except the inclusion of the semi-skilled worker.

At this point, Figure 4.1, which started the conversation, needs updating (see Figure 4.4).

AUTOMATION

Automation has many forms: mechanical devices, electronic devices and monitoring equipment, etc. By itself, automation does not guarantee food safety, but it can reduce risk with proper equipment maintenance and workforce training.

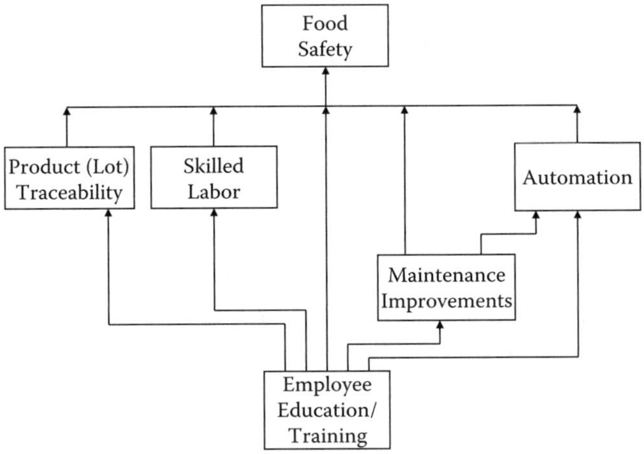

FIGURE 4.5
The effective use of automation is directly linked to maintenance, which in turn is linked to education and training. (From APICS International Conference & Expo, October 21–23, 2007, Denver, CO. With permission.)

Often a process is automated to save money—not to increase food safety—and automation can be expensive. But anything that reduces human error in the processing of food is good, and both directly and indirectly contributes to the pursuit of food safety. Figure 4.5 highlights the connections between effective automation and equipment maintenance and employee training.

SYSTEMS INTEGRATION

Discussing systems integration and supply chain integration is like discussing what came first—the chicken or the egg? System integration is generally thought of as integrating all the information flow processes for:

- Recipe/formulation design and management
- Customer order management
- Production management
- Supply management
- Quality management, including regulatory record keeping
- Human resources, including employee training plans and status
- Financial systems such as accounts receivable (AR), accounts payable (AP), etc.

Within the boundaries of one company, using information technology and managing everything from one common database is a means to achieving system integration.

If a company operates from numerous unconnected computer spreadsheets, uses undocumented manual processes, and just has integrated financial systems, i.e., AR to the general ledger (G/L), it doesn't have (internal) systems integration. The analogy of a chain only being as strong as its weakest link applies here. Systems integration is a prerequisite to being a reliable partner within a supply chain, particularly when passing critical food safety data down the supply chain.

If a company is operating without the benefit of complete systems integration, what can it do? Investigate and learn about Enterprise Resource Planning (ERP). ERP is an integrated planning, controlling, and centralized data repository based on concepts, processes, and techniques that support best practices, and has evolved over a forty-year period. In the early days of its evolution, the computer software industry understood the opportunity to apply computer software to emerging best practices. At the time, the marriage of the evolutionary predecessors of ERP and computer technology was considered disruptive, giving significant competitive advantage to those who had implemented it and effectively used it in a disciplined manner. During the early days, large companies were the primary beneficiaries of ERP, particularly those who built devices of some sort.

Over time industry-specific variants of ERP were developed, including those that addressed the special needs of food and nutritional supplements manufacturers. The good news is that once most of the large companies committed to ERP and the market was saturated, the opportunities for growth for software developers shifted to smaller manufacturers—and with it more appropriate costs of ownership. The version of ERP designed for food manufacturing is often referred to today as process batch ERP. More good news: There are lots of process batch ERP choices for the small- to medium-sized company. Figure 4.6 is a simplified illustration of process batch ERP. Process batch ERP will be examined in more detail later in this book.

SUPPLY CHAIN INTEGRATION

Almost every industry is "chasing" supply chain integration, not just food and supplements manufacturers. The current conventional wisdom is that

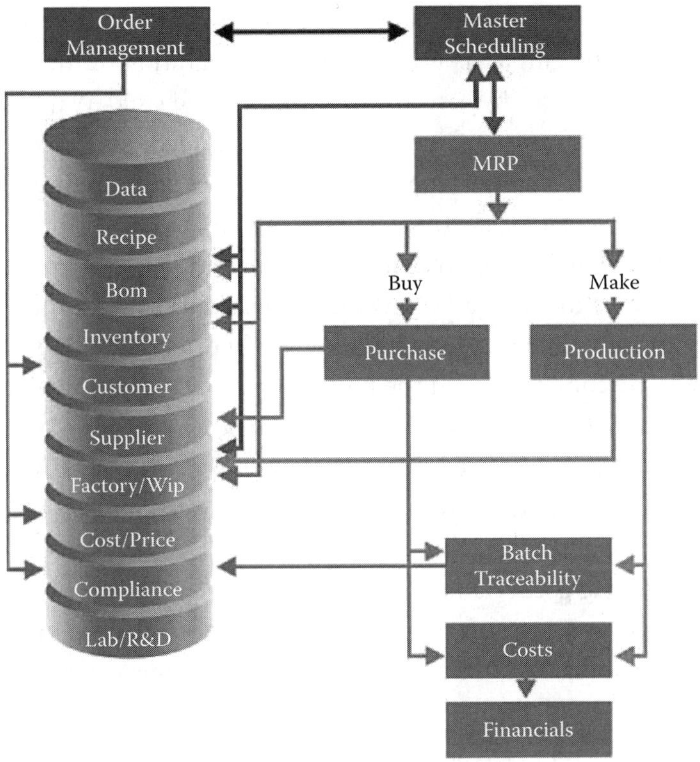

FIGURE 4.6
A simplified view of batch-process ERP and its integrated subsystems. (From APICS International Conference & Expo, October 21–23, 2007, Denver, CO. With permission.)

companies don't compete against other companies, supply chains compete against each other. What supply chains a company chooses to be part of, or is allowed to be part of by its customers, greatly influences a company's prosperity.

The unfortunate thing is within many industries there are often differing definitions of what supply-chain management is and what level of integration is needed. The reason for this is that complex supply chains or long supply chains for many companies are relatively new. Why is that? About twenty years ago the concept that an organization has one or just a few core competencies, and most of them then had a vertically integrated infrastructure including many things that a company didn't really do well, started to gain acceptance. The push was to shed anything that wasn't within a company's core competencies and subcontract it out to suppliers.

Up to that point, the emphasis was to vertically integrate all data and processes within the four walls of the enterprise and over the forty-year period Enterprise Resource Planning (ERP) evolved. It supported procurement, but the emphasis was on vertical integration of the enterprise. Almost in lockstep with this shift in thinking, the Cold War ended and very rapidly the sources of supply greatly expanded, and a truly global economy started to emerge. Supply chains became more complicated, and with more participating partners. In the lingo of supply chain there were more upstream and downstream nodes. Upstream nodes are suppliers. The result was a move to a horizontal business model, and the planning and control systems being used were vertically oriented. There were (still are) major "disconnects," and the best example of this is noted in a *Wall Street Journal* (a U.S. business newspaper) article on the lack of visibility to upstream suppliers on what was going on with the ultimate customer, the supply chain master. This profiled supply chain was global. The financial impact on many trusted supply chain partners mentioned in the article was dire, creating great economic hardship.

OK, how does a lack of visibility within a supply chain impact food safety? In Chapter 3, an audit by the U.S. Inspector General was mentioned, which found that the majority of products examined for a complete lot trace history failed. Without an accurate and complete lot trace record throughout the supply chain, food safety is compromised. Supply chain integration brings with it not only supply and demand information, but also the required accurate lot trace data.

On the positive side, there are resources to improve the situation. Two nonprofit professional associations stand out:

- The Supply Chain Council (SCC) is the organization that developed the Supply Chain Operations Reference (SCOR) model.
- APICS is the association for supply chain and operations excellence.

Both reach out to industry for proven and emerging best practices, codify (document) the practices, create courseware to broadcast the practices, organize forums for corporate and practitioner education and networking, and have rigorous practitioner certification processes for those practices. The SCC has the certification P/SCOR, and APICS has two: the Certified in Production and Inventory Management (CPIM) and the Certified Supply Chain Professional (CSCP).

MORE ABOUT SCOR

> SCOR® helps manage a common set of business problems through a standardized language, standardized metrics, and common business practices which accelerate business change and improve performance. The SCOR model provides a unique framework that links business process, metrics, best practices and technology features into a unified structure to support communication among supply chain partners and to improve the effectiveness of supply chain management and related supply chain improvement activities.
>
> **—http://supply-chain.org/**

The building block of SCOR is simple but powerful, and for each company in a supply chain, it standardizes the documentation of these processes. The first step is to document each contributor in the supply chain across the following:

- Plan
- Source
- Make
- Deliver
- Return (may not apply to food and nutritional supplements in most cases)

While this makes sense, how many suppliers have taken the time to understand their suppliers' plan/source/make/deliver processes and those of their suppliers' suppliers?

The SCOR process flow is shown in Figure 4.7.

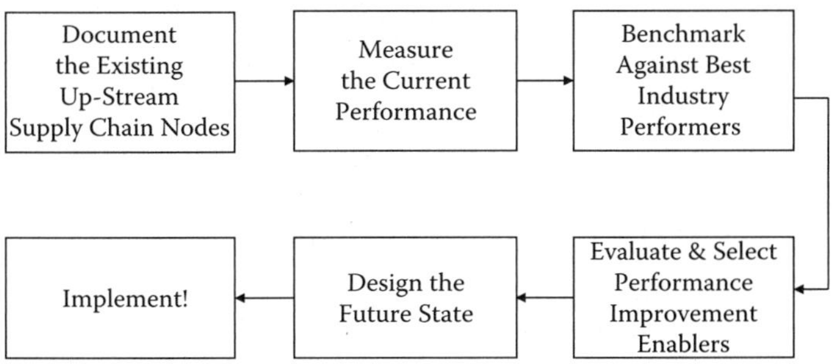

FIGURE 4.7
The process steps to implement the SCOR methodology.

A BRIEF OVERVIEW OF APICS

APICS has a multitiered approach with emphasis on increasing the competence of individual operations and supply chain practitioners and also corporations through education on proven best practices. Over the years it has developed standard courseware to teach best practices, developed a dictionary of standard terms, and established a framework to understand all the best practices within an enterprise and across a supply chain, and provide guidance on career planning for both the practitioner and the corporate employee development staff.

Certification is also a focus of APICS, and it has two major certifications that are highly respected, Certified in Production and Inventory Management (CPIM) and Certified Supply Chain Professional (CSCP), both of which are internationally recognized.

APICS uses multiple channels to deliver education:

- Through chapters geographically dispersed around the world. In North America, there are over two hundred chapters with volunteer management and instructors.
- Collaboration with universities.
- Web-based courses.
- Collaboration with other professional associations.
- Major multitopic conferences.

Because there are cost-effective resources available to help equip companies to improve supply chain integration, there is no reason for lack of progress. Figure 4.8 “plugs in” systems integration and supply chain integration into what is needed from the top ten list to achieve food safety.

There is more complexity than is reflected in Figure 4.8 in that supply chain integration is required for incoming raw materials, then systems integration as the company produces product, followed by supply chain integration as the next processor adds value to the product until a final product gets into the hands of the consumer. That is where the value of the SCOR Framework comes into play. Understanding what actually happens as product passes through to its destination to the consumer is critical to food safety.

There are two issues that haven’t been analyzed and discussed:

- Increasing capacity
- Improving equipment or batch changeover

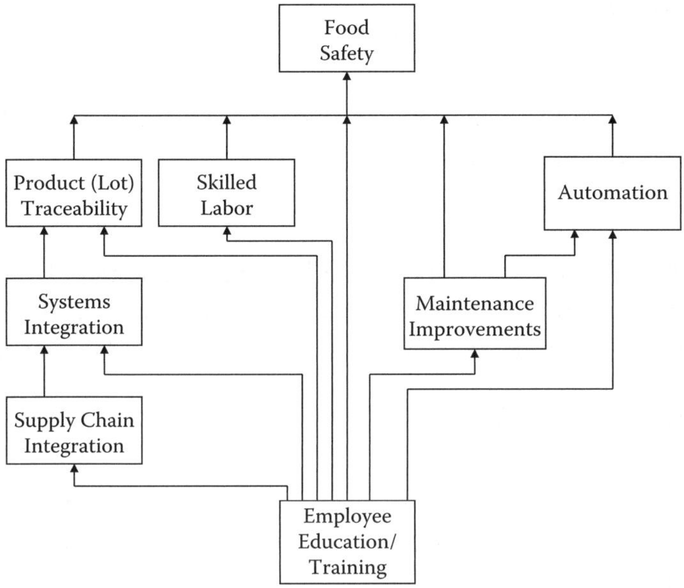

FIGURE 4.8
Supply-chain integration is a prerequisite for systems integration, and in turn product (lot) traceability.

In reality both are beneficiaries of improvements in one or more of the seven business issues that result in improved food safety. For example, effective use of ERP can improve capacity utilization. SMED can improve equipment changeover time and the quality of the change result.

What is very clear from the following exercise is the importance of education and training to food safety. Education and training influence the performance over seven critical areas in the top ten list just examined that contributes to food safety and improved operating performance.

In the beginning of this chapter its objectives were defined as helping the reader understand the connectivity between the issues and putting together an appropriate strategy and creating a priority ranking for attacking the issues, by answering the questions:

- Where does one start the improvement process, and what is the appropriate sequence of improvement projects?
- It is important to think differently. What is the context? Are we addressing a symptom or a root cause? Are there enabling processes required to mitigate the issue?

QUIZ

4.1. The purpose of this chapter was to only examine each of the top ten food industry issues.
☐ True ☐ False

4.2. Since the passage of the record-keeping requirement for lot traceability in the USA Bioterrorism Act, the food industry has been very successful in developing complete and accurate lot trace records.
☐ True ☐ False

4.3. The record-keeping tools used today for lot traceability are generally the same for both large and small-sized companies.
☐ True ☐ False

4.4. Today information technology (IT) systems are cost-effective for large-sized companies only.
☐ True ☐ False

4.5. Skilled labor is readily available and at low cost.
☐ True ☐ False

4.6. There is a blueprint and process for developing a skilled and semiskilled workforce.
☐ True ☐ False

4.7. Improving equipment maintenance can be accomplished by what four approaches or techniques?

4.8. Verbal instructions to a semiskilled worker suffice as adequate employee training.
☐ True ☐ False

4.9. There is a direct correlation between employee education and training and maintenance improvements and effective use of automation.
☐ True ☐ False

4.10. ERP was designed around the goal of systems integration.
☐ True ☐ False

4.11. ERP has been enhanced for food and nutritional supplement manufacturing.
☐ True ☐ False

4.12. There is a linkage between the philosophy that a company should focus on its core competence and the current supply chain thinking.
☐ True ☐ False

4.13. There is consensus on what supply-chain management is.
☐ True ☐ False

4.14. Supply chains are less complicated today than twenty years ago.
☐ True ☐ False

4.15. Lack of visibility throughout many supply chains is a significant problem.
☐ True ☐ False

4.16. Most companies have a detailed understanding of the supply chains they participate in from field to fork.
☐ True ☐ False

4.17. There are cost-effective resources to help a company improve its systems and supply chain integration based on best practices.
☐ True ☐ False

The answers to these questions can be found in the Appendix along with definitions, website addresses, and recommended reading.

5

Defined Processes, Procedures, and the Importance of Data Accuracy

This chapter discusses the importance of well-written standard operating procedures (SOP) and why Hazard Analysis Critical Control Point (HACCP) thinking makes sense even if it is not required for the products being manufactured. Following the discussion on SOP and HACCP, there is a more detailed examination of the importance and organization of quarantine or hold areas, options on organizing the receiving and warehousing areas in a way that meets cGMP requirements and reduces errors (mistake-proofing), as well as lowering indirect costs. This chapter also introduces a Lean thinking principle of the visual workplace.

This chapter also describes the implied requirements of satisfying the lot traceability record-keeping obligation. It establishes the linkage between a disciplined approach to record keeping and the management of processes and their positive relationship in providing a foundation for superior inventory and supply-chain management.

Considerable attention is focused on data quality or data accuracy because it is one of the keys to achieving or exceeding product safety regulatory compliance and effective use of planning and control systems that support best practices and lead to improved organization performance. Questions on what needs to be accurate and why, how accurate, and how this level of data accuracy can be achieved across the organization are discussed in detail. Included are illustrations and a game that demonstrates the impact that poor data accuracy can have and its ramifications, showing the importance of data accuracy.

cGMP, SSOP, SOP, AND HACCP

Both the food and the nutritional supplement manufacturer have the benefit of a framework on which to build capabilities to ensure the manufacture of safe food. That framework is cGMP. With more and more mass production of food, safety issues developed. Corrupted food impacted larger numbers of consumers. The issues had to be addressed and guidelines had to be established. Over time cGMP for the food industry evolved as the body of knowledge on food science and safety increased. For many years both food and nutritional supplements manufacturing were guided by a shared and common cGMP set. Today the supplements industry operates under a different set of cGMP, which are based on lessons learned from experience with food cGMP and designed in consultation with the industry; they are based on current nutritional science and address the differences between the product types. Any proven standard or guideline when implemented provides an advantage. A company doesn't have to reinvent the wheel, and there is ready access to "how to" educational sources and expert guidance. Another benefit of cGMP is their flexibility; they tell you *what* must be done, not *how* to do it. A company is free to develop its own "how to do it" as long as it satisfies the cGMP. This is reflected in a company's sanitation standard operating procedures (SSOPs).

Most of a company's workforce performs work per one or more SSOPs and isn't required to have a deep knowledge of cGMP. It helps to have someone understand why work is being performed a certain way and the consequences for food safety, but it isn't absolutely required as long as the SSOPs are clearly written, comprehensive, and thorough and the worker is properly trained to perform the task. The responsibility to develop SSOPs is with the company's regulatory compliance specialist, quality specialist, and production management working as a team. Once again, there are many cost-effective educational resources on how to develop an effective SSOP, as is the case of understanding cGMP.

Not all SOPs in a company deal with food safety, but all contribute to process predictability, including the administrative processes and information flow. Most individuals assume that processes are conducted in a consistent and managed fashion, and everyone understands what's going on. The reality is that many processes aren't performed the same way every time because they haven't been analyzed, documented, and people have not been trained on them. Assuming others know without verification seems to be a human trait. Tribal knowledge is used throughout many operations as the tool for making products.

If work isn't performed the same way each time, it can't be improved. Since food manufacturing isn't a high-profit-margin business, improving processes can mean the difference between making a profit or incurring a loss.

Those companies that have committed to Lean thinking (continuous improvement) use a technique called standard work. The purpose of standard work is performing work the same way each time.

There is a companion technique called kaizen that is focused on improving work performed in a specific targeted area for improvement. When the targeted improvement is realized, the new standard work is then documented, workers are trained on it, and it is implemented. Both standard work and kaizen and a high-level process analysis technique called value stream mapping will be examined in greater detail in subsequent chapters of this book.

HACCP is another figurative gift from the U.S. regulatory authorities. It is so highly regarded it has been integrated with ISO 9001 (quality focused) into the ISO standard 22000. The International Organization for Standardization (ISO) is globally accepted and often has advantages over national standards for those exporting. Although HACCP is mandatory for only certain food products in the United States, it is food safety focused and has a powerful underlying analytical logic. Many companies have adopted it even though it isn't required for them because it suggests a disciplined approach to food safety and is a marketing differentiator. It looks at risk in a systematic manner, but most importantly, it is teachable and implementable. It requires:

- Determining critical control points (CCPs)
- Using science-based acceptable (critical) limits on each CCP
- Determining and designing a plan how the risk will be minimized
- Defining what documentation must be maintained to confirm conformance to the HACCP plan
- Designing an appropriate response when the product processed through a CCP exceeds the critical limit
- Assigning responsibility for each aspect of performing to the HACCP plan
- Assigning responsibility to the trained individual who has plan ownership and authority to sign off on all HACCP-related documents

The key thought is that CGMP, SSOP, SOP, and HACCP create, if rigorously implemented and adhered to, a reliable foundation for producing safe food if:

- There is a logical design of workflow from beginning to end, starting with the procurement function through the manufacturing process and ending with shipping of product to the customer
- All work areas are organized to be efficient, work procedures are performed as designed, and the data generated executing them are accurate, complete, and communicated in a timely manner
- Cost-effective technology has been deployed to minimize data entry and the potential for errors and workers trained on their use
- The receiving department area is physically organized to visually provide the status of incoming materials, especially hold (quarantine) areas
- The warehouse/stockroom layout supports not only efficient put-away of incoming and issuing of outgoing (to production) materials, but also segregation of material by lot numbers that facilitate a first in, first out scheme
- It has been confirmed that all individuals who record information, store product, or issue product have a proven attitude for accuracy in counting and ability to be disciplined in executing procedures
- There is an effective lot control system in place for tracking materials and finished products

The seven points mentioned are focused on receiving, storing, issuing materials/ingredients to production, and shipping the physical product, all linked with an information-rich lot control record-keeping system. But there are other contributors to ensuring food safety:

- Accurate recipes/formulations and processing instructions
- Everyone understanding the insidious and cumulative impact of inaccurate data and the required level of accuracy across the organization

Let's examine each of the nine requirements in more detail.

A LOGICAL WORKFLOW

The food industry has a history of entrepreneurship, much more so than other industries. Very small-sized companies have grown to small companies and progressed to medium-sized companies. There have been many corporate failures within the industry, but many more to replace them. Small companies have small facilities, a small workforce, and simple approaches to managing. Those that survive innovate to success with a great product concept. The U.S. FDA estimates that there are in excess of one hundred thousand food manufacturers in the United States alone. Most of those companies have less than two hundred employees. Include companies in the European Union, Asia Pacific, India, and South America and the number of small- to medium-sized food companies is staggering.

The successful small food company operates on a pragmatic basis. Access to money to finance operations is difficult, so everything is scaled to cash flow, small and manageable, and day-to-day thinking prevails. With success come larger facilities, much needed capital investments, and increases in workforce size. Success brings with it more compromises in the growth planning process. Life in "the trenches," which may require wearing more than one management hat (multiple responsibilities), isn't easy. In this instance, designing an end-to-end efficient process flow is often neglected, and the skills of the workforce to deal with the dynamic of a larger company underestimated.

But eventually growth levels off, competitive pressures build, and it's time to reevaluate how the company operates. Looking at how procurement, inventory management, product manufacturing, and support services, i.e., recipe development, etc., flow from end to end is critical in understanding what to focus on for improvement. Improving or rationalizing the process flows isn't just about moving the business to the next level of economic growth, it's also about quality and food safety. One of the standards of measuring quality within an organization is ISO 9001. Think of it as the *Consumer Report* on a company's quality management. If a company doesn't understand and document process flows, there is no ISO 9001 certification in its future.

The good news! There are simple techniques to document and understand process flows. The first is a spaghetti diagram. If anyone has eaten a plate of spaghetti, the noodles don't lie down in neat lines but are confused and very visual. A spaghetti diagram is a visual line representing the movement of work as it is performed and moves through physical locations. A spaghetti diagram isn't a Rorschach test to determine sanity. But it will generally show that there is an opportunity for improvement. A spaghetti diagram is a wakeup call to action, and Figure 5.1 is a visual example.

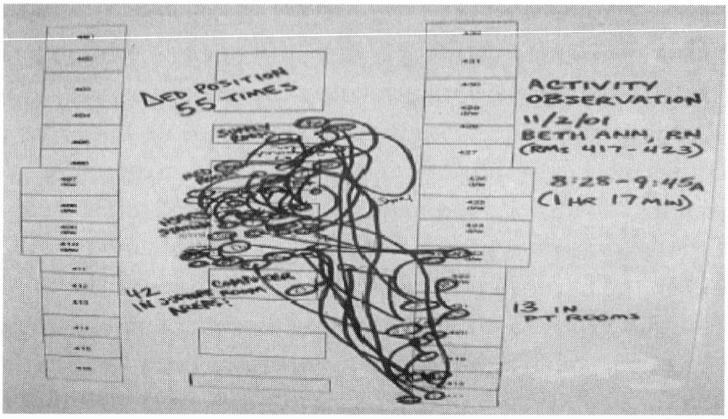

FIGURE 5.1
Spaghetti diagram: before.

One senior food industry executive recounted how a spaghetti diagram on one worker revealed that he walked over seven miles per day to perform simple, related tasks. While walking the individual didn't add value to producing product but was getting a good cardiovascular workout! Clearly there was an opportunity for improving the process flow, and the company did that by engaging in a dialogue with the employee, analyzing the work performed, designing the ideal state, relocating some machinery, confirming with the employee that the new process flow made sense, and finally confirming by observation that it did.

Spaghetti diagramming is the foundation for a more sophisticated approach to analyzing and designing an efficient workflow called value stream mapping (VSM). VSM adds data to the spaghetti diagram. It looks at the amount of time spent performing value-added work in each step of a process flow and how much time is spent idle or waiting for value-adding work to be performed in each step. Also documented is the information flow that supports each step in the process flow. Figure 5.2 shows an example of a VSM for the manufacture of pasta products.

The example VSM indicates that over 41 days of lead time were required in the current state, with only approximately 1.5 days of actual value-added activity applied! Carrying excessive inventory is expensive, not only including finished goods but also work in progress (WIP).

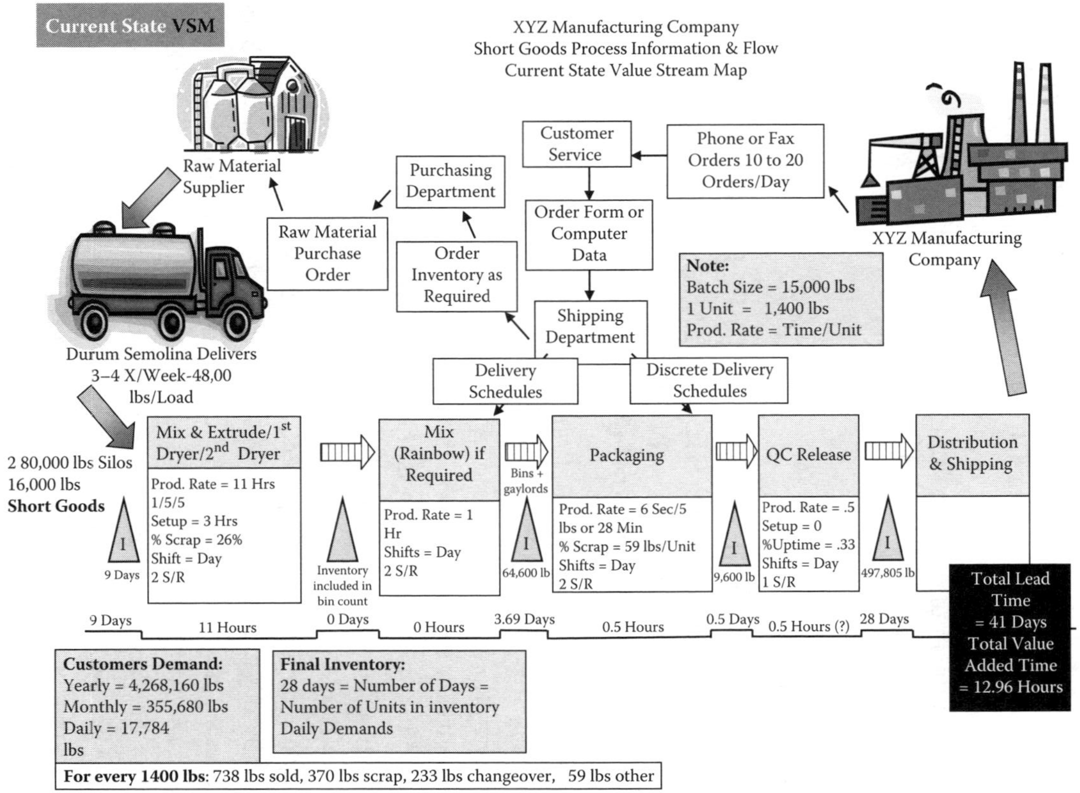

FIGURE 5.2
Current-state VSM. (From Ron Calame, California Manufacturing Technology Consulting® (CMTC), Torrance, CA. With permission.)

That 39.5 days of wait is pure waste, some of which can't be eliminated, but much of it can. But why from an economic standpoint is this relevant? The cost of (excessive) inventory is calculated by most perceptive cost analysts at eighteen percent plus the cost of money.

While the economics of fixing convoluted workflows is compelling, their impact on food safety is more urgent. There is a law called Murphy's law. It states that if anything can go wrong, it will. Illogical, inefficient workflows and periods of non-value adding invite Mr. Murphy to the party.

Value stream mapping will be examined along with the "how to fix it" techniques in later chapters of this book.

WORK AREAS ARE ORGANIZED TO BE EFFICIENT, WORK IS PERFORMED AS DESIGNED, AND DATA ARE ACCURATE, COMPLETE, AND COMMUNICATED IN A TIMELY MANNER

Most companies think in terms of improving productivity once an operational or task step is started. Earlier the problem of idle or wait time in between operations was discussed, which increases lead time and WIP inventory investment, both of which are pure waste and consume profit margins. Value stream mapping is the start of eliminating this waste by highlighting what work areas need to be reorganized. Another opportunity is to mistake-proof reporting of information as work is performed. Prompt and accurate recording of data is a cornerstone of food safety. For many companies making transactional inputs, such as what lot numbers were processed into a production batch, at what day and time, and who performed the work, is an afterthought in designing work areas. Often workers have to access reporting mechanisms that are located far from the point where work is performed or have to aggregate information and report it infrequently as a batch input to remain efficient.

The problem with a work area that makes it difficult to promptly record data is that people often forget to record information when there is a time lapse because recording it is arduous and it is human nature to make mistakes or forget. As Murphy's law states, if anything can go wrong, it will, and this acknowledged law will eventually prevail. There were always negatives associated with poor design in recording lot-related data within a work area, but that primarily reduced profit margins. Since 9/11 we've been made aware that the scope of risk is far greater.

Integrating the ability of recording status data as work is performed isn't really difficult, it just requires understanding the value of doing so, a focus on doing it, and mistake-proofing how it's done. The next section examines how cost-effective technology can support minimizing the "hit" to productivity and enhancing data accuracy.

COST-EFFECTIVE TECHNOLOGY HAS BEEN DEPLOYED TO ELIMINATE DATA ENTRY ERRORS

Ten years ago information technology-oriented equipment such as bar code printers and readers, wireless routers, mobile devices, and "ruggedized" personal computers/workstations were mostly used by larger firms because of the relatively high costs. Large companies have historically been early adopters of useful technology and have in practical terms functioned as a research and development (R&D) lab proving what worked and what didn't. Early adopters generally gain a competitive edge with the use of new useful technologies. As with any new technology, the original sponsoring customer segment (large companies) eventually buys enough to satisfy its needs. Using the language of professional marketers, it's saturated. Then two things happen; because the suppliers of the equipment have recovered investments in R&D and capital equipment, they start to look for new market segments, such as smaller firms. They leverage the success they've had with larger firms in marketing to the smaller firms, and because many of the significant costs no longer exist, they lower the price and the process of commoditization starts. Because of this the smaller organization can now afford to invest in formerly leading-edge, technology-centric equipment. The benefit:

- Bar coding minimizes human data entry. Research has shown that anything over six digits or characters in an identifier has a significant error percentage when entered by a human. Identifiers include raw material number, batch number, worker number, parent batch number, etc. Since these devices can be mobile and wireless, they can be placed at the point of work and reduce administrative overhead in reporting. From a food safety record-keeping perspective the information is accurate and captured.

- Wireless technology provides the freedom to allow the worker to report from just about anywhere in the facility. It has become so pervasive in our day-to-day lives outside of manufacturing that we take it for granted. If it's cost-effective for a coffee shop as a benefit to its customers, it must be cost-effective. If individuals are using within their homes wireless printers, PC wireless keyboards, and a wireless computer mouse, it obviously is a cost-effective commodity. Anything that makes it easier to input information at the point of work and in real time makes food safety record keeping easier. While bar coding and wireless don't necessarily need to be combined, it does make sense to do so.
- While Enterprise Resource Planning (ERP) isn't equipment, it does integrate with all the latest technologies, i.e., bar coding and wireless devices. The benefit relative to data accuracy derived from ERP is that it operates off of one centralized database and will do a real-time editing of all information sent to it and maintain a log of transactions for investigation of anomalies that need to be addressed. ERP will be addressed in detail in Chapter 6.

The use of any device requires some level of training. In the case of using bar code readers and wireless at the worker level, very little is required but shouldn't be taken for granted. Everyone needs to know the importance of using these devices properly. Supporting this capability behind the scene with technical staff does require more education and training, but nothing overwhelming. Remember the "how to" support aspect has been commoditized for use by smaller companies.

Depending on the role and responsibility of the individual involved, using ERP potentially requires more education on concepts and the internal logic of various calculations, as well as how information flows from customer order entry through to shipment.

In the past, education or training was relatively expensive. With today's technology, tutorials can be embedded in many systems at the point where work is performed and in multiple languages, often with diagrams or video-based instruction for the visual learner. Modern ERP systems are excellent examples of this. If embedding the instructions for its use in the device system isn't practical, suppliers often provide documentation and instruction over the World Wide Web (Internet).

Even if a supplier doesn't provide affordable predesigned training, a company can create its own with user-friendly technology such as YouTube, a web-based video service where simple short tutorials can be made using a video camera and microphone, posted on YouTube, and easily accessed with a web browser seven days a week, twenty-four hours a day.

In summary, the equipment and systems to minimize data errors to support food safety record-keeping requirements *and* reduce administrative costs of entering data *and* increase worker productivity are now practical and affordable and supported by cost-effective worker training.

THE RECEIVING DEPARTMENT AREA IS PHYSICALLY ORGANIZED TO PROVIDE A VISUAL STATUS

A concept that has growing acceptance in manufacturing is the visual workplace. The goal of this concept is to organize work areas so that anyone walking through an area understands what is going on at a semidetailed level without having to ask questions or go to a computer screen to get answers. A simple status board is often used with preprinted cards similar to kanban cards that can be moved from column to column to reflect status. An example status board for receiving and receiving inspection could include:

- How many incoming purchase order receipts are planned or scheduled to be processed today by receiving (a numerical sum).
- The items and purchase order numbers scheduled to be received today (detailed information on the preprinted card).
- The high-priority items established by planning with the preferred sequence (detailed information on the preprinted card, color coded, and sequenced).
- The items physically received but not processed by receiving (indicated with a checkmark).
- The actual cumulative count for the day against the goal for receiving until shift end (a numerical sum).
- How many items or orders are planned or scheduled to be processed today by receiving inspection (numerical sum).

- Purchase order receipts processed by receiving into inspection and the preprinted cards moved into the inspection column of the visual display board, including high-priority items (detailed information on the preprinted card).
- If an item requires lab testing the preprinted card is moved into the column for lab testing and a due date is noted based on prenegotiated lead times with the lab. Preprinted cards are sequenced by due date.
- If there are issues, such as no Certificate of Analysis (C of A), failed lab test, no lot trace number, etc., the preprinted card is moved to the issues column for corrective action.

This simple status board described above would look like that shown in Figure 5.3.

A key consideration is the location of the status board itself. It must be easily accessed and seen by both receiving and receiving inspection staff. Doing a spaghetti diagram and mini-value stream map can help in deciding its exact location.

Date: Today

Receiving	Inspection	Lab Testing	Issues
Goal: 25 Actual: 5 Item #/PO #	Goal: 25 Actual: 6 Item #/ PO #/ Lot Trace #/ COA	Item #/ PO #/ Lot Trace #/ COA/ Due Date	Item #/ PO #/ Lot Trace #/ COA/ Due Date
√ Modified Food Starch PO # 0914	Carrot Puree, PO # 0902/ LT# 69931	Crab, PO # 0875 LT # 69922	Tuna, PO # 0871 LT # 69911
√ Potassium Chloride PO # 0903	Cayenne, PO # 0904/ LT# 69930		
Onion Powder PO # 0996	Modified Potato Starch, PO # 0980, Lt# 69927		
Corn Protein PO # 0905	Sea Salt, PO # 0972, LT# 69925		
Calcium Chloride PO # 0901	Garlic Powder, PO # 0877 LT# 69928		
Soy Protein Isolate PO # 0899	Maltodextrin, PO # 0997 LT# 69932		

High Priority Item Needed for Production Scheduled this Week

FIGURE 5.3
Receiving and receiving inspection status board.

There is an opportunity for improving the visual status board. Have the preprinted cards bar coded, and give workers bar-coded employee ID badges and bar code scanners, mobile scanners if possible. Scanning the cards could update whatever is being utilized as a centralized data repository. A batch process ERP system is the ideal repository, but other approaches can be put together.

An obstacle to achieving the visual workplace is insufficient space and overcrowding. Part of this can be overcome by carefully scheduling the incoming shipments and estimating the workload beforehand. Often, but not always, the constraint will be QA inspection. Scheduling overtime when peaks are anticipated is an initial remedy, but one that should be used sparingly when other measures are implemented. The goal is to have very little WIP left over at the end of the workday, with the exception of shipments received in the late afternoon, those out for lab testing, and those requiring special disposition.

Estimating the workload of a shipment receipt and inspection required isn't difficult; it just requires an approximation. Suppliers can also be required to ship to arrive no earlier than X number of days before the due date of the shipment. When the amount of raw material is large or very bulky, often suppliers will store it for their customer locally and release shipments as required. When arrangements like this are made, it is in everyone's best interest to share benefits: supplier visibility into expected demands, reduction in inventory carrying costs, etc. It has to be win–win.

While factory space is expensive—real estate costs, heating and cooling, maintenance, etc.—the cost of a confused and overcrowded work area is substantially more; it's just hidden from our conventional cost accounting thinking, as many other things are. More space is better than less space, and it doesn't always mean increasing square footage. Going vertical can be good even if it's only temporary.

What space provides is the opportunity to segregate and organize in a visual way. Something as simple as color-coded lines bounding an area that represents a certain status, such as:

- Awaiting the start of the shipment receiving process
- Shipments missing: documentation, or don't have a lot trace number, or not the declared quantity, or the product is damaged, etc.

- Receiving processes complete and available for incoming inspection
- Quarantined for issues or lab testing
- OK for movement into inventory or issue to production

Organization and clarity translate into data accuracy, velocity through the organization, and improved food safety record accuracy.

Please note that companies who have success in their Lean journey (continuous improvement) often report reductions in the space required in their march to the visual workplace.

There is a third dimension to getting organized to achieve the visual workplace in receiving and receiving inspection. That dimension is being measured on performance that contributes to the bottom line, food safety, and showcases individual and team achievements. The receiving and receiving inspection status board is focused on today. Physical organization makes it easier to keep everything easily understood and focused. Continuous measurement is about the past, the now, and the future with no excuses. The measurements must be carefully thought out because poor choices can be destructive, while rational measures elevate and create performance. Measures need to be tailored to each business environment. A few possible weekly, monthly, and quarterly measures include:

- Planned receipt processing transactions goal against actual (week to date, weekly, monthly, quarterly)
- Receiving department team attendance target versus actual (week to date, weekly, monthly, quarterly)
- Receiving department team overtime versus actual—important during peak seasonal periods (variable)
- 6S inspection results versus targets—discussed later in this book (weekly, monthly, quarterly)
- Process improvement training techniques, training (or review), planned versus actual (quarterly and annually)
- Pounds or pallets processed per man-hour

The goals and actual results to data are best presented visually in the form of, but not limited to, bar charts, pie charts, and trend charts, and maintained in a highly visible area within the workspace. Whatever resonates with the worker team is what should be used.

THE WAREHOUSE/STOCKROOM LAYOUT SUPPORTS EFFICIENT MATERIAL HANDLING AND ALSO SEGREGATION OF MATERIAL BY LOT NUMBERS

In the ideal situation, a stockroom or warehouse would be very small because the company had a high number of inventory turns. Many successful companies who have been involved in continuous improvement (Lean) activities have gotten very close to eliminating traditional stockrooms and warehouses by synchronizing supply and demand, achieving a high inventory turns performance, and storing inventory at the point of use because the amount of inventory stored is minimal and has a small physical footprint because of the exceptional inventory turns performance. The philosophy is that every time you touch inventory but don't add value by consuming it in a timely fashion you create waste (add costs). This thinking is very food safety centric in that less of everything is easier to manage and data accuracy easier to maintain, and in the event of a product recall, the rapid turnover of product should in theory make the recall more focused on the offending lot(s). *Precision* is the operative word. While this is a great future-state goal, the reality is that most companies would be best served initially by having a larger stockroom/warehouse space to keep raw materials, intermediates, and finished goods segregated by lot number with first in, first out processing easy to maintain. Space makes it easy to make everything visual and improve mistake-proofing. Fortunately it can be reclaimed and eliminated over time with improvements in supply chain and inventory management while maintaining the benefits of ample space.

There are numerous schemes to organizing a stockroom or warehouse. The various choices have been documented, best practices codified, and education and training courses developed and marketed. A major consideration is the ease of putting inventory into the stockroom and pulling it for issue to customer and production orders. Unlike companies that make little red wagons, those that are operating under cGMP have guidelines on some aspects of stockroom layout. But what works in one environment doesn't necessarily work in another. Below are a few examples.

A distribution company's purpose is to be a provider supplying inventory items quickly and eliminate the need to deal with multiple suppliers. The benefit of a distribution company is the reduction of administrative costs and often reduced landed costs resulting from favorable pricing secured because of volume procurements. Because of high volumes, a typical distributor and large-sized manufacturer can invest in extensive order put-away and picking automation, unlike most small- to medium-sized manufacturers. A distributor will often implement a warehouse management system (WMS), a specialized integrated system. Typical picking techniques are piece picking or picker-to-part method and sorting systems method, both of which utilize automation to pick parts for orders.

Small- to medium-sized manufacturers often don't have the access to money to finance high levels of automation in the warehouse and use picking methods such:

- Piece picking/picker-to-part method, in which the stockroom picker moves to each part or ingredient to fill one specific order.
- Zone picking method, in which each order picker is assigned a specific zone and will only fill orders within this physical zone. This is a good method if raw materials are clustered by product family within the stockroom.
- Wave picking method, in which the picker will move to collect the materials required for several orders.

Choosing the method to store incoming and kitting of materials for a customer or production order will influence the stockroom/warehouse layout. The key to organizing the physical stockroom and the order put-away and order picking process is analysis and common sense. An example: One company that utilized a manufacturing cell concept and the traditional approach of stockroom organization by item number reorganized storage into mini-stockrooms that minimized the distance material had to travel from storage to a cell, and transit time. Slightly more low-cost labor, but significantly more velocity in material movement.

Another important consideration is material segregation by product and lot trace number that is easy to see. For segregation that supports first in, first out issuing of materials by lot number, think again in terms of the visual workplace. Organizing visually often requires more storage space. The benefits are fewer mistakes in issuing the preferred lots and improved data accuracy, a requirement to satisfy accurate record keeping in support of effective lot traceability. There is a temporary downside; space costs money, but if a company commits to being Lean, the inventory turns will increase and the need for space will decline and can be redeployed to more profitable uses.

CONFIRM THAT ALL INDIVIDUALS WHO RECORD INFORMATION, STORE OR ISSUE PRODUCT HAVE A PROVEN APTITUDE FOR ACCURACY AND ABILITY TO PERFORM WORK IN A DISCIPLINED MANNER

No matter how much thought and effort is put into organizing a workspace, it can be laid low (brought down) by those supervising and others doing the day-to-day work if they don't have certain aptitudes. If one accepts that food safety and operational excellence are supported by structure and orderliness, process thinking, discipline, the ability to accurately see and assess, and count accurately, this highlights the need for those in receiving, receiving inspection, and on the stockroom team to have these attributes. It shouldn't be assumed that all individuals have them. Education and training are important, but without the required aptitudes it's an uphill battle. Not having the required aptitudes isn't a defect, it just means an individual would be better utilized in a different role. Who wants a jet engine mechanic overhauling a jet engine who doesn't have mechanical skills? Or a financial accountant who is mathematically challenged? The qualities required for each team member in the material receiving, inspection, and inventory control process flow can be observed through simple testing and observation.

The seven points examined are focused on receiving, storing, and issuing materials/ingredients to production, then shipping the physical product and accurately recording lot traceability data at each step.

But there are other contributors to ensuring food safety:

- Accurate recipes/formulations and processing instructions
- Everyone understanding the insidious and cumulative impact of inaccurate data and the required level of accuracy across the organization

ACCURATE RECIPES/FORMULATIONS AND PROCESSING INSTRUCTIONS

There is an expression regarding data accuracy that transcends industry type: "garbage in, garbage out" (GIGO). Data accuracy is one of the cornerstones of effective food safety, and the recipe or formulation is the baseline document for everything that follows: the HACCP plan, product costing, planning, raw material procurement, QA, and manufacturing. It's also a requirement for achieving world-class operations and supply-chain management. A recipe brings with it other considerations when selling the product: ingredient and nutritional statements and special declarations such as low sodium, low fat, organic, high fiber, etc.

Unfortunately, recipes and formulations that do have inaccuracies often go undetected because workers correct what they know is incorrect or deviate from the official ingredient list for a variety of other reasons. When companies do audit a sizable sample of recipes against what is actually done, they find these errors, and often it is eye opening to the auditors *if* they fully understand the implications, which will be discussed below.

The opportunity for data errors increases when a company utilizes multiple disconnected computerized spreadsheets as the database to run the organization, which is not uncommon in small companies.

Another vulnerable official document is processing instructions. An example would be a change in the equipment utilized that could change some aspects of the product manufactured. The number of processing steps does influence the risk of error; the more steps, the higher the likelihood of potential errors.

It's easy to dismiss small data errors in the official recipe and processing instructions, but these can compound and be amplified into a whole lot of risk and confusion, which will be examined next.

THE INSIDIOUS AND CUMULATIVE IMPACT OF INACCURATE DATA

Sometimes people think accuracy is not important, and approximation is sufficient. This might be true in some cases, but inaccurate data can be a silent organizational cancer. Sometimes "close enough," even if seemingly minor, can be damaging. An example that requires some mathematical calculations follows:

> A ship, a freighter with commercial goods, is leaving the U.S. port of New York for a transatlantic trip. The destination port is Saint Peter Port in the United Kingdom. Nothing tricky about the trip, it's pretty much a straight line with a northeast trajectory and without many potential complications other than weather. During the start of the journey the navigation

> equipment is found to be inaccurate and in need of calibration. The captain is a seasoned and skilled individual and decides to continue the journey. Finally the ship arrives at a port. Before the reader jumps to a conclusion consider a 2° error in the context of 360° degrees and it doesn't seem that much. Only, the ship arrives in Ile De Groix, France, missing its planned destination port by 163 miles, or 262 km! This is not a small deviation. Ever run a 163-mile marathon, or walk 163 miles with only brief rest stops and sleep periods, or pay for a 163-mile cab ride? Or if you're the captain telling the ships owners that their expensive transportation vehicle (the ship) would be delayed with increased crew and fuel expenses? How about the importers waiting for their goods? Talk about being in a difficult situation.

The situation is hypothetical but the math isn't (Figure 5.4). The point of this example is that seemingly small data errors can create big problems not only in operational performance but also in food safety regulatory compliance. Recalling many additional lots in the event of a recall because of a lack of data accuracy, no matter how small the error, has significant economic consequences.

Another way of looking at the impact of inaccurate data is to look at the confusion it causes. The damage done by inaccurate data can be deceptive, even with a high percentage of accuracy. An example would be accuracy of a specific type of data at ninety percent. Let's call this the ninety percent

Given:
Distance from A to B = AB = 3286.56
Distance from A to C = AC = 3400.42
Distance from A to D = AD = Cos(2)* 3286.56 = 3284.56
Distance from B to D = BD = Sin(2)* 3286.56 = 114.70
DC = AC AD = 3400.42 3284.56 = 115.86
Square of BC = Square of BD + Square of DC
Thus BC = 163.03 Miles

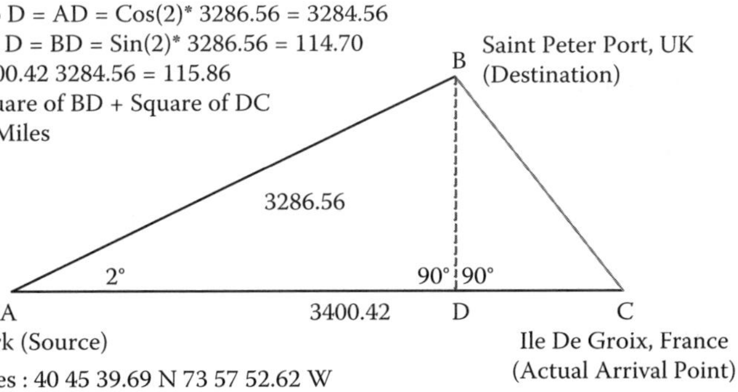

A Point Coordinates : 40 45 39.69 N 73 57 52.62 W
B Point Coordinates : 50 14 49.42 N 4 22 28.22 W
C Point Coordinates : 48 14 43.08 N 3 17 31.38 W
Source of Data : Google Earth

FIGURE 5.4
Example of the calculation details on the impact of the data inaccuracy. (Software images provided by BatchMaster Software, Inc. With permission.)

data accuracy dilemma. Ninety percent sounds good until you consider the amplification effect across the following:

- 90% recipe accuracy ×
- 90% bill of material accuracy ×
- 90% work instruction accuracy ×
- 90% inventory record accuracy ×
- 90% inventory issuing accuracy ×
- 90% factory reporting accuracy

90% × 90% × 90% × 90% × 90% × 90% = 53.14%, appropriately labeled the confusion factor. What seemed to a good level of accuracy cascades as work progresses through the organization and becomes a figurative tsunami of inaccuracy. How do companies currently survive this? By workers making adjustments to keep product moving forward; this creates hidden waste of resources and increases risk and undermines the integrity of food safety.

What is the appropriate level of accuracy? Industries outside of food and nutritional supplements have long recognized, for example, that inventory record accuracy should be at a minimum of ninety-five percent with an ideal target of ninety-nine percent.

The good news is there are techniques to achieve the ideal data accuracy targets.

SUMMARY

In summary, this chapter discussed defined procedures and processes, and the importance of the following was examined:

- How SSOPs, SOPs, and HACCP provide a solid foundation to be built upon
- The implied requirements of effective lot traceability
- The importance of logical workflows
- Techniques to fix convoluted workflows
- The importance of organized work areas and inventory storage areas and guidance on how to improve them

- How cost-effective technologies can help eliminate data errors and a brief overview of several different and useful technologies
- The importance of education and training and tips on delivering it on a cost-effective basis
- How to improve the physical organization of receiving and create a visual workplace
- Suggestions on organizing the warehouse/stockroom in support of efficiency *and* food safety
- The importance of employee aptitude in achieving and maintaining accurate data records
- Why the recipe needs to be accurate
- The hidden cumulative impact of inaccurate data

QUIZ

5.1. Education and training on operational best practices, the role each person plays in maintaining food safety, and the effective use of information technology are cost-prohibitive for the small manufacturer.

☐ True ☐ False

5.2. Adhering to food safety regulatory requirements negatively impacts operating efficiency.

☐ True ☐ False

5.3. Achieving a lot trace record accuracy of ninety percent is considered acceptable.

☐ True ☐ False

5.4. There is no relationship between the physical layout of a work area and accurate record keeping.

☐ True ☐ False

5.5. Name two techniques that can help in understanding existing workflows and what areas need to be focused on to improve them.

5.6. What three cost-effective information-based technologies can help reduce data errors?

5.7. The visual workplace keeps everyone up to date on the status of work in process within a work area.
☐ True ☐ False

5.8. The visual work area is difficult to establish.
☐ True ☐ False

5.9. What two obstacles need to be removed to set up a visual work area?

5.10. Keeping lot numbers segregated in a warehouse or stockroom is optional.
☐ True ☐ False

5.11. The extra space needed to organize work areas to minimize errors is an ongoing perpetual cost.
☐ True ☐ False

5.12. The obligation to maintain data accuracy is limited to receiving, inspection (QA), and warehousing and production staff.
☐ True ☐ False

5.13. All individuals have the aptitude for accurately recording information.
☐ True ☐ False

5.14. The consequences for small data errors are negligible.
☐ True ☐ False

The answers to these questions can be found in the Appendix along with definitions, website addresses, and recommended reading.

6

Enterprise Resource Planning (ERP) and Supply-Chain Management

This chapter examines Enterprise Resource Planning (ERP), a complete and proven end-to-end planning and control system that supports both effective food safety compliance and operating excellence. The key points examined are:

- Why ERP?
- A definition.
- In the beginning (the dark ages).
- ERP's history and evolution.
- The historical return on investment (ROI) from effective use of ERP.
- The impact of data accuracy on the effective use of ERP.
- Enhancements for other manufacturing industries.
- The differences between standard ERP and process batch ERP.
- Is ERP compatible with Lean?
- The emergence of the current supply chain philosophy.
- Familiarity breeds contempt; however, ERP is still critical to competitiveness today.
- How to get up to speed quickly on the best practices embedded in ERP.

WHY ERP?

There are numerous reasons: It can make a significant contribution to the bottom line, it is proven and the "how to" in order to prepare a workforce to effectively use it is known, and guidance and help to put it into all business processes is readily available. Perhaps even more important is the fact that many of the requirements for its successful use are shared with best practices in food safety, in particular its emphasis on:

- Creating workforce awareness of cause-and-effect relationships
- Understanding process flows
- Using proven techniques
- Careful planning
- Achieving and maintaining data accuracy
- Discipline in executing the plan(s)
- Free flowing and timely communications
- A commitment to professionalism

Another reason is that your customers just might prefer suppliers who have good controls over their operations and are effective in using ERP.

ERP didn't just recently spring into existence; it's neither an invention nor an information technology gadget. It has evolved over forty years through trial and error and has leveraged information technology to put in place best practices, from forecasting customer orders to collecting money for shipments made. Manufacturing was the first sector of our economy to have a comprehensive planning and control framework: ERP. Since then other industries have selectively taken techniques from this discipline and applied them with success.

All the different systems within ERP and how they interact together can be easily depicted graphically, and as the chapter progresses there will be illustrations to help the reader visualize how it all fits together. About halfway through the chapter a variant of ERP, process batch ERP, which evolved in the mid-1980s, is examined in detail. Process batch ERP shares the underlying logic of traditional ERP but has capabilities unique to food and nutraceutical manufacturing.

ERP—A Definition

Historically the use of the acronym ERP has been greatly abused, with it being often applied by the uninformed user to any subsystem or function within the enterprise. Fortunately there is a *de facto* standard for definitions related to supply chain and operations management called the *APICS Dictionary*. This is a comprehensive dictionary that has eliminated word confusion and improved communications within companies committed to improving operational performance. APICS is a nonprofit professional association and considered the thought leader for best practices in supply-chain and operations management. The definition for ERP from the APICS dictionary is:

> Enterprise Resource Planning (ERP)—Framework for organizing, defining, and standardizing the business processes necessary to effectively plan and control an organization so the organization can use its internal knowledge to seek external advantage.

There is a companion publication from APICS that for want of a better description is an encyclopedia on all the concepts and techniques that support best practices in operations and supply-chain management. It's called the Operations Management Body of Knowledge (OMBOK) Framework. Both volumes warrant a place on anyone's desk (or computer) who is in pursuit of performance improvement.

In the Beginning (the Dark Ages)

Some years ago there was a stimulating educational TV program called *Connections*. It connected obscure historical developments that eventually resulted in significant breakthroughs in concepts, science, products, and social movements. Think in terms of connecting the dots and the domino effect in combination. ERP has its roots in the connections paradigm. The next few paragraphs are in that vein.

After World War II each major industrial nation was heavily influenced by its then existing infrastructure and manufacturing capacity. This forced each country into a specific strategy on how to manage its resources. Strategies ultimately create concepts and techniques on how to best satisfy the strategy. One way of thinking about this concept is called Maslow's Hierarchy of Needs (McLeod, 2007). Maslow pointed out that the most basic needs were safety, shelter, and food. Once satisfied, a person or organization moved to a higher level of need.

During the war the United States had built up immense industrial capacity without any catastrophic damage to this capacity. Combining the war rationing and the preceding economic depression of the 1930s created huge unsatisfied demand for goods. The average U.S. citizen was living on a shoestring budget for a long time. Capacity to produce product wasn't a problem, there was lots of it. The emphasis shifted to managing inventory and satisfying customer demand. Getting the right part to the right place at the right time was the challenge. The planning and controls tools needed to do this had not yet been developed. The inventory planning tool of choice then was reorder point. Reorder point looks at near-term history and current lead time and suggests you commit to more production or procurement at X number of weeks. It isn't connected to calculated demand. During this time period computing technology wasn't practical or affordable for the majority of manufacturers. The end result was companies surrounded themselves with a figurative wall of expensive inventory.

Changing the Focus and Example

By the conclusion of the war, Germany's infrastructure was badly damaged, its labor force shattered, and its industrial capacity greatly reduced. This meant rebuilding capacity requiring significant investments to do so. The focus was on utilizing every machine and human resource. Producing excess inventory wasn't a concern since in aggregate there wasn't much that didn't get consumed; the supply situation was very tight. From this emerged systems that collectively were called production planning system (PPS) that were designed to leverage capacity. Postwar Germany also strategically positioned itself to be an exporter of highly engineered goods. Generally speaking, the management of inventory has become a focus in Germany only in the last twenty-five years.

Japan is another country that evolved a different competitive strategy after the war. Its infrastructure and industrial capacity were also badly damaged. Japan has historically been a resource-poor nation, at least in terms of industrialization. As mentioned earlier in this book, Japan can only provide forty percent of its caloric needs domestically and must import the rest. Despite this, it has been tremendously successful because it understood its challenges and responded.

If you have few resources you have to make more with less. Without access to domestically extracted natural resources such as oil and natural gas, you're at a competitive disadvantage and must be more efficient than the competition in all aspects of manufacturing and supply chain processes. Success also requires differentiation from competitors. Also, limited resources in all forms point to the need to collaborate with other companies and leverage joint resources. Putting it all together, Japan committed to continuous improvement (CI) that combined a focus on quality with elimination of all waste. The words used today to reflect all this are *Lean* and, prior to that, *just-in-time* (JIT) manufacturing.

Back to the discussion of the United States and its concern about inventory. Without inventory that matches customer demand, the product can't be made. If you can't make it, you can't ship it. Unshipped customer orders don't create revenue. No revenue, no inbound money. The response to this dilemma in the dark ages was to build a wall of inventory to deal with our lack of systems to calculate and match demand and supply. Unfortunately surplus inventory isn't cheap, is incredibility expensive. Experts estimate that surplus inventory—that is, inventory that can't generate cash now, *excluding* the cost of money for borrowing—is approximately eighteen percent per annum. Add in the cost of borrowing money and the number is crushing. Way back then, this profit improvement opportunity got the brain cells of a number of innovative problem solvers rubbing, and the search for a technique to precisely calculate and balance demand and supply began. It was the beginning of the evolution of ERP.

ERP'S HISTORY AND EVOLUTION

This section details the evolution of ERP, including business drivers that drove its evolution and how each system in it works to support:

- Recipe management
- Maintaining product processing instructions
- Quality assurance
- Documenting and managing departmental capacity
- The development of corporate planning
- Customer order management
- Planning production and purchasing orders
- Scheduling production and purchasing orders
- Maintaining manufacturing and purchase order status
- Managing by exception
- Centralizing and maintaining data
- Management by performance metrics
- Development and capture of detailed financial data

Material Requirements Planning (MRP)

The evolution of ERP started in the 1970s, and the business driver was getting rid of the "wall of inventory" that negatively impacted operating costs. A then (and still) disruptive technology, computerized information technology, had just barely crossed the threshold of affordability for large corporations that also had large and costly surplus inventories. The secret to reducing surplus inventory was precisely matching supply with demand requirements, and that required many calculations to be done quickly. A number of thought leaders developed a system called material requirements planning (MRP). This cadre of thought leaders included individuals such as Joseph Orlicky, Oliver Wight, George Plossl, and Richard Ling.

The logic of MRP is straightforward and could technically be done without computing assistance, but the number of calculations, opportunity for errors in calculating, and time required to build a plan made the manual approach unattractive; by the time you finished the plan it was obsolete. MRP requires specific data on:

- Recipes, also known as the bill of material (BOM) or "parent" that defines each ingredient required to make the parent and how much of each ingredient is required
- What the lead time is to make the recipe (BOM/parent)
- What the inventory status is for each recipe, what's in the storeroom/warehouse
- What is on order (authorized) for each recipe, the quantities and due dates
- How many firm orders for the recipe are committed to and for what date and for how many
- How many orders, both internal and external (intermediates or customer orders), are planned or provisionally planned and their due dates
- Specific planning information on each ingredient to make the recipe, replenishment lead times, minimum and maximum quantities that can be ordered at one time, costs, etc.

Figures 6.1 to 6.3 are simplistic graphics of a recipe in a BOM-like structure and MRP data requirements and outputs. As with any new approach, there were successes and failures. Why did some companies succeed and others fail? The successful companies' approaches to implementing and

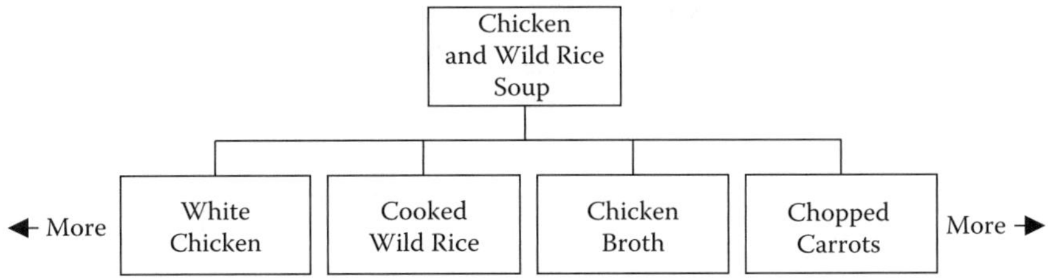

FIGURE 6.1
A single-level recipe (BOM).

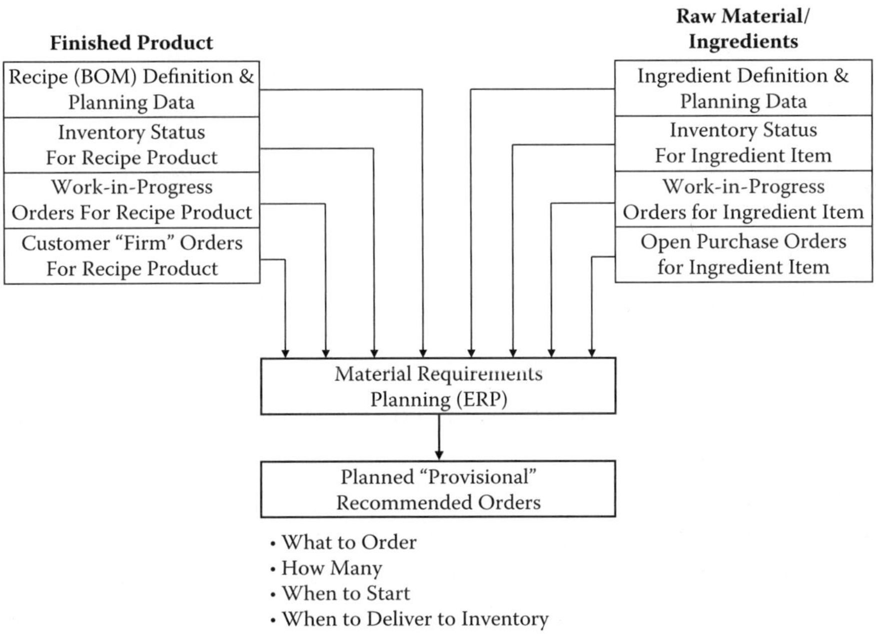

FIGURE 6.2
A visual representation of MRP calculations and the outputs.

using MRP were examined, as were the companies that failed. From these examinations there were lessons learned and MRP became part of something more comprehensive called closed-loop MRP.

Closed-Loop MRP

There were four main lessons learned from the successes and failures in getting results from MRP:

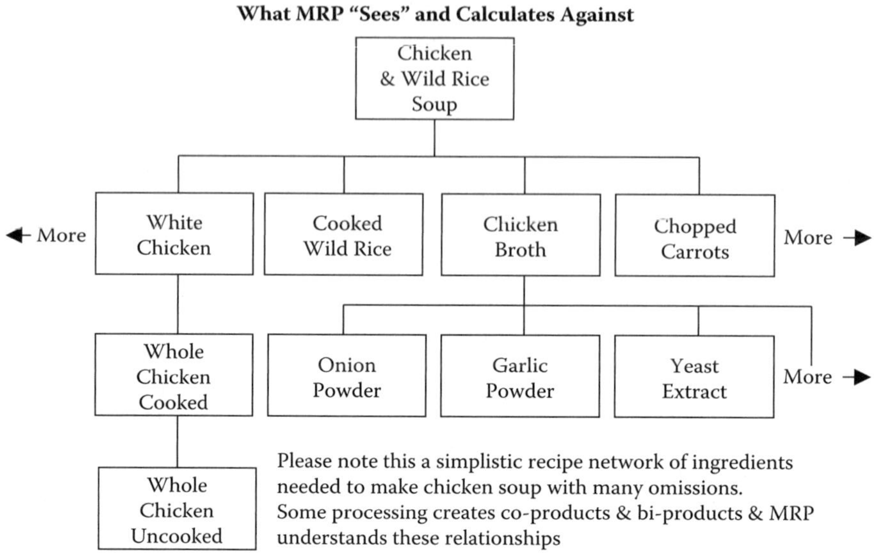

FIGURE 6.3
What MRP "sees" and calculates against.

- Data accuracy was critically important.
- Changing habits were difficult and education and training requirements underestimated.
- MRP, because of its precise calculations, could be nervous with indiscriminate changes in due dates of orders, quantities required, etc., creating many recommendations to take corrective action.
- A planning system needed a companion system that managed the execution of the plan.

Data Accuracy

In the figurative dark ages, when reorder point was the planning system of choice, everyone knew that there would be lots of surplus inventory and high levels of expediting to deal with part/ingredient shortages. Accuracy wasn't given much thought since chaos was the norm. The importance of having accurate inventory records and accuracy in all other data was not generally recognized. MRP changed that. Its design drove it to balance supply with demand and, if possible, leave zero surplus inventory. The successful MRP pioneers understood the importance of data accuracy; those that failed did not. Chapter 5 briefly examined the impact of inaccurate data and the confusion it causes within the organization.

What is important to recognize is that the data accuracy required by MRP is at the same level of accuracy required to maintain records in support of mandated lot traceability.

Techniques to assist in getting accurate inventory record data have been developed, such as cycle counting and control group cycle counting. Cycle counting has several variants, but the recommended first step is control group cycle counting, because it's a fast way to identify the root cause of the inventory record inaccuracy through the auditing of most of your inventory processes with a small number of items. The goal is to fix the processes of:

- Picking inventory from predetermined storage locations
- Putting inventory into predetermined storage locations
- Recording all relevant data on the movement of inventory

Or identify who requires additional training if these individuals are the root cause of inaccurate data being generated. If an error in the amount in the inventory record is found, or the location is incorrect, or lot trace numbers don't match, the cycle counter investigates to determine the root cause of the error. This individual must have the aptitude to count correctly, but also be something of a combination of Sherlock Holmes and a bloodhound that follows the scent.

The items in the control group are counted on a daily to weekly basis, and continue to repeat this process for all selected controlled items until no errors have occurred for ten consecutive times, which is an indicator that the processes are fixed, and all involved employees trained.

At this point, using another cycle counting method is appropriate and the 80/20 rule to stratify the counting plan into three categories, A, B, and C, is often used. Those that qualify as A items collectively constitute eighty percent of the total inventory value. The A group might also include long-lead-time items or those considered strategic. These are counted more frequently than items in the B grouping, and the items in the B grouping more frequently than those in C.

The Role of Education and Training in Implementing Organizational Change

Most organizations resist change. Getting outside of a comfort zone creates pressure. Pressure is the enemy of inertia; inertia is comforting to the average person. The antidote or countermeasure to inertia is education followed by training. That is the lesson learned from those successful MRP users.

The first component of an education curriculum is to communicate what's in it for each person, not just the corporation. More job security because there is less waste and confusion. Personal gratification because each person is working for (or with) a winner who is committed to excellence and each contributes to its success. Following this, the second component is creating familiarity with the new way in progressive stages.

Using history as an example, the first layer of learning about the Roman Empire was that it had military dominance and excellence in military strategy and tactics. The second layer was their skills in structural innovations and infrastructural building. The final stage was learning about their ethics and morals, including strategic mistakes made.

Learning in layers seems to be a great way of avoiding rejection of change. There is a concept called acceptance time. People need to understand and marinate in the planned new situation.

Getting it right means not applying a one-size-fits-all scheme. Each discipline or profession in a company has a different culture. This means the education plan for each has to be "tuned" and carefully planned. Also, effective education and training require not overeducating or undereducating. They must be delivered without a lapse in time, before the new way of thinking can be used by the employee/student. Unused new knowledge dissipates quickly.

It's not clear if those early successful companies thought through the importance of education and training to implementing a different way of thinking or just let their intuition guide them.

Fortunately all the education on best practices required to plan and manage inventory has been predeveloped by associations such as APICS, and the science of how to deliver effective education by the American Society for Training and Development (ASTD).

The Root Cause of Nervous MRP

In a typical manufacturing organization, there are hundreds, if not thousands, of changes daily to order quantities, items unexpectedly scrapped, changed priorities, etc. MRP will precisely recalculate a new set of recommended actions based on this nervous situation. An appropriate analogy for MRP is that it is like the human body, which is capable of intricate movements but must be directed. What made MRP nervous was that it lacked a figurative brain, a place where higher-level decisions were made to ensure stability, a system more aligned with shipping schedules that were controlled and managed with great care. What emerged to fit this need was master production scheduling (MPS). The deployment of MPS was the major contributor to the success of pioneering MRP users.

Master Production Scheduling (MPS)

There are several MPS styles, but they share a common goal of reducing the number of items that must be master scheduled. In the minds of many, the product shipped is what the MPS directly manages. For many companies this can mean hundreds, if not thousands, of end item shippable products. If there are too many items to manage, then the quality of decision making on balancing supply and demand and maintaining a stable schedule is difficult, if not impossible.

Reducing the number of items to manage requires analysis and thinking outside the box. One technique is to focus on product families and to find where the maximum value is stored in the smallest number of items and start thinking in terms of modularity (Figure 6.4).

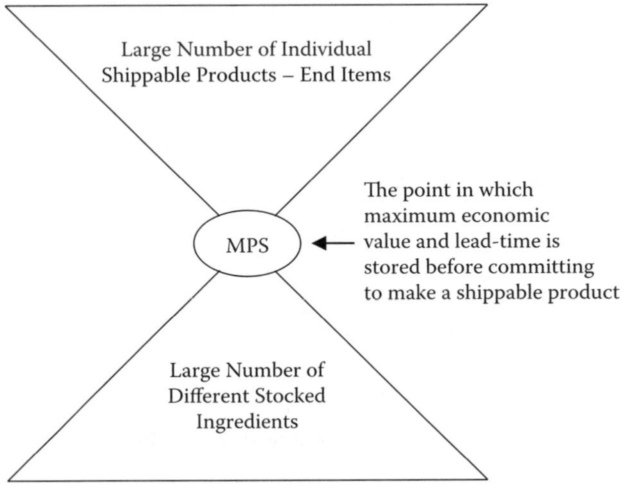

FIGURE 6.4
MPS—the supply–demand management "sweet spot."

The advantage of product modularity, if it does exist in a company's product portfolio, is demonstrated by the simple luncheon menu in Figure 6.5. If you look at the number of options they total forty-six:

- Proteins—15
- Breads—6
- Condiments—5
- Salads—9
- Beverages—11

The math: 15 × 6 × 5 × 9 × 11 = 44,550 end items to manage! In today's consumer-driven economy in which they want more choices, it's reasonable to think that there would be additional choices offered to maintain

Protein Choices

☐ Beef patty	☐ Roast beef	☐ Chicken salad	☐ Pastrami (hot)
☐ Salami	☐ Corned beef (hot)	☐ Ham	☐ Ham and Swiss
☐ Cheese	☐ Turkey (sliced)	☐ Reuben	☐ Shrimp salad
☐ Tuna	☐ Roast pork	☐ Egg salad	

Bread Choices

☐ Wheat	☐ White	☐ Rye	☐ Pumpernickel
☐ Roll	☐ Sub		

Condiments

☐ Mustard	☐ Mayonnaise	☐ Butter	☐ Lettuce
☐ Tomato	☐ None		

Salads

☐ Carrot	☐ Potato	☐ German potato	☐ Macaroni
☐ Coleslaw	☐ Tossed green	☐ House special	☐ None

Beverages

☐ Tea	☐ Pepsi	☐ Coke	☐ Diet Pepsi
☐ Diet Coke	☐ Dr. Pepper	☐ Vernor's ginger ale	☐ Coffee
☐ Milk	☐ Iced tea	☐ None	

Number of Options

Number of Configurable Lunches

FIGURE 6.5
Typical customer-configurable luncheon menu.

or increase revenues. Even with a modest addition of choices, let's say one for each choice category, the number of end items to manage increases to 80,640!!

Forecasting accurately for this many end items is difficult, if not impossible. But getting the numbers right for forty-six items is easier and doable because there is time to investigate and understand customer demand patterns.

In this example it might be possible to reduce the risk of stock shortages and reduce inventory further by looking at demand at a higher level of aggregation, for example, for the protein choice. Perhaps organizing the fifteen choices into meat, poultry, fish, ethnic/regional, and vegan would offer advantages because the larger a logical grouping is, the easier it is to forecast accurately. With this level of aggregation there might be volume discounts from raw material suppliers. What about each choice within a choice category? How are shortages avoided? By slightly overplanning each option based on historical data, current trends, and product promotion strategies.

Beyond the advantage of managing far fewer end items, the modular approach offers another advantage, the opportunity to postpone committing to an end-item configuration until there is more certainty that it can be sold. Using the luncheon menu example, most of the value, lead time, and flexibility is stored in the modular choices.

One constraint to using the modular approach (if appropriate) is the traditional recipe/formulation or bill of material (BOM). Its purpose is to define what is needed and how much of each ingredient is required. Change anything in the recipe and you have a new end item. This doesn't simplify planning and scheduling or managing product data. The approach by many organizations is to create planning BOMs that are used strictly for planning. Often the recipe-formulating team will also adopt this approach and the two will merge.

Understanding what the best MPS approach should be is a mix of proven technique, thinking outside the box, and art. The definitive text on understanding product modularity is a book titled *Mass Customization* by Joseph Pine (see the appendix). *Mass Customization* addresses the blindness caused by familiarity phenomenon.

Not all companies will have the benefit of food product modularity. Some really are selling specialized capacity in the form of equipment or workforce capability. One of the things that MPS was designed to do is test the feasibility of the plan before committing to it, which requires making sure capacity is available to execute the plan or, conversely, that expensive resources are being utilized by the plan.

The subsystem that evolved to do that is rough-cut capacity planning (RCCP). Rough-cut was designed to only look at critical resources and make sure the plan was feasible and achievable, and if not, adjustments would be made. Adjustments could include not only changing dates or quantities, but shifting work to other equipment, scheduling overtime, or temporary out-sourcing.

Rough-cut is the first step in a two-step capacity planning process; it confirms the viability of the MPS before it is authorized and released. The second step is capacity requirements planning (CRP), which is a detailed look at capacity based on the output of MRP (Figure 6.6).

Executing the Plan

The final system needed to extract the full potential of MRP was a detailed scheduling and order status system for manufacturing operations. In the late 1970s most companies still tried to make as much product internally versus out-sourcing, which is favored today. Today many companies have embraced the concept of operating off their core competence and subcontracting everything outside of it. Nevertheless, a manufacturing execution system is still critical to success. It manages the gap or activities that MRP established between the start and finish date. There are a number

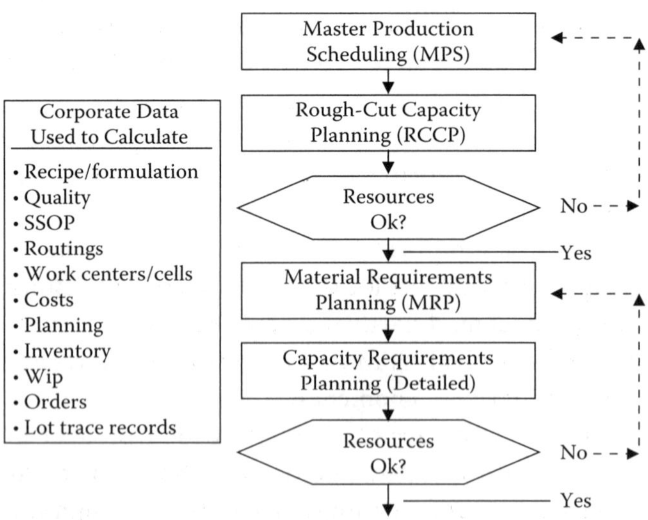

FIGURE 6.6
Making sure the plan is feasible compared to capacity constraints in a multi-step "go-no-go" planning process.

of different names used to describe an order execution system: shop floor control (SFC), production activity control (PAC) and manufacturing execution system (MES). The objective of each is to schedule and monitor progress for each operation required to complete an order. There is even provision for manufacturing cells that combine a number of processing steps and one-piece flow, a technique favored by those adhering to the just-in-time (JIT) or Lean philosophies and methodologies.

The data needed for manufacturing execution systems are shared or utilized by a number of different corporate users:

- Industrial and manufacturing engineers (IE & ME): Establish the data to standardize what equipment to use, how work should be performed, and the time standards for each operational step. This definition is done for each processed ingredient, intermediate, and finished product. They also define a work area's capabilities and capacities, often in collaboration with manufacturing management. By defining how much work-in-progress inventory is acceptable for each department and creating a workday calendar, they create the basis for factory capacity requirements planning (CRP), detailed scheduling by operation, and alerts for off-schedule orders, and even the planned lead times for MRP. With the addition of work center costs and cost targets for the center the foundation for product costing and performance metrics is laid down.
- Marketing and finance: Product costing utilizes the data established by industrial engineering as an input to final product costing.
- Planning: Capacity requirement planning (CRP) uses the same data to calculate the capacity demand generated by the MRP plan against stated capacity.
- Planning: MRP lead times can be calculated by using these data. Some closed-loop MRP systems can calculate different lead times based on batch size.
- Management and finance: Performance measurement and cost variances are calculated off the same baseline data.

The advantages of one data repository with integrated data that are shared and reused by many different departments started to be recognized. Contrast this with the status quo for thousands of small- to medium-sized food and nutraceutical manufacturers today using disconnected spreadsheets to manage and keep regulatory compliance records. Operating with many unconnected electronic spreadsheets creates issues of data integrity and interdepartment coordination in an environment that requires precise actions and record keeping to ensure food safety. It also adds significantly more workload to an organization's staff; it is a burden and hardship.

A quick mini-summary:

- Data accuracy is critically important.
- Education and training can change habits.
- MRP is an important system but needs complementary systems working with it.
- MPS is the brain that directs the body and is management's handle on the organization.
- A planning system needs an execution system.

When the reasons for some of the early MRP failures were analyzed and understood, and the reasons for those who succeeded were broadcast to the manufacturing community, many companies committed to using this new road map called closed-loop MRP and succeeded. There were still some failures, but these were generally attributed to management incompetence.

Success breeds success and creates a very positive dynamic. With a larger pool of innovators who wanted to build onto their success they pushed ERP into its next evolutionary step, MRP2.

MRP2

MRP2 incorporated at least seven innovations or logical improvements over closed-loop MRP:

- A higher-level plan created by senior management called sales and operations planning (S&OP) that created boundaries for MPS.
- Significant enhancements to customer order management capabilities.
- The integration of financial ledgers accounts receivable (AR), accounts payable (AP), general ledger (GL), and fixed assets (FA) with closed-loop MRP.
- Significant enhancements to support the purchasing function.
- Additional methods of planning, in particular recognition of just-in-time (JIT) thinking. This was recognition that one planning technique didn't fit all. The concept of hybrid planning tools took root.

- Enhancements to support other styles of manufacturing, i.e., recipe or formula-based products such as food, pharmaceutical, and aerospace and defense (A&D).
- Manufacturing didn't stop at the boundaries of North America, and translations into other languages of the body of knowledge that supported MRP2 started.

Sales and Operations Planning (S&OP)

S&OP is designed to bring together not only sales and operations executives but also engineering, quality, regulatory compliance, finance, and supply-chain management to agree on a unified operating plan with support as needed from other functions. Its focus is on product families and the use of critical assets. The intent was to create an actionable plan in contrast to the then typical annual business plan that was often too vague to execute against. Early and successful MPS users came close to S&OP thinking with the use of multilevel MPS. Since the amount of detail was significantly less than required for MPS processing, using computers to do what if analysis was quick and practical.

What S&OP isn't about is just balancing the numbers between sales, manufacturing, supply chain partners, and finance, but a commitment to interdepartmental collaboration, recognition that the whole is greater than the sum of its parts, and that optimizing something at the expense of other important processes can be damaging. Figure 6.7 incorporates the S&OP into the planning process.

Customer Order Management

Closed-loop MRP was a manufacturing planning and control system that with maturity started to evolve into a business system. Business starts with customers, and customers want to work with suppliers that are easy to do business with, build quality products that are price competitive, are reliable, understand their shared history, are quick to respond, and provide accurate information. All of this requires organizing data in a systematic way and having quick access to it. Much of the required data already resided in computerized closed-loop MRP, and extending and enhancing it made sense. The term often used back then was the *order management system*, which continued to evolve.

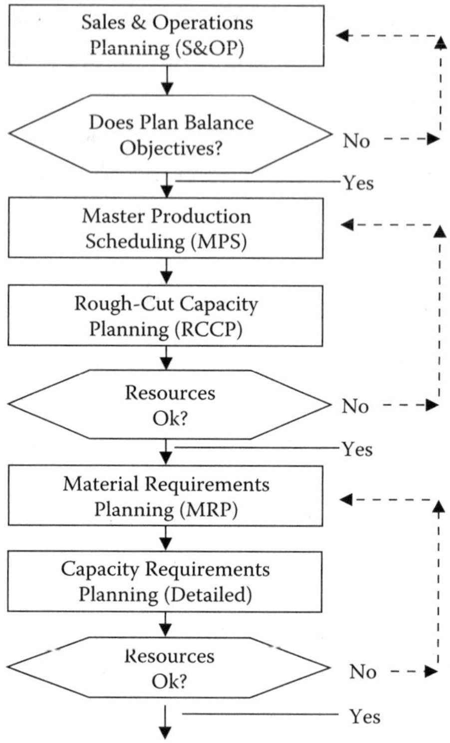

FIGURE 6.7
A top-down planning process that confirms feasibility and creates a "contract" between all the corporate disciplines.

During this time, many forward-thinking companies started to understand that being customer centric made business sense and systems that supported the relationship between the customer and them needed to be designed with the customer's needs in mind, and not just internal efficiencies in working with the customer. A well-regarded book titled *Service America* (see Appendix) talked about this change and chronicled many success stories. This new perspective influenced the development of customer order management best practices.

Integration with the Financial Ledgers

Every activity that occurs in a manufacturing company has money figurative attached to it. Book a sale, buy a part, use equipment, make a part, store a part, manage activities, etc. All the raw information needed to feed the financial ledgers is generated in the customer management and manufacturing planning and control systems. Cost management, accounts payable, general ledger segments, and accounts receivable use transaction data from these systems. The financial ledgers are set up with rules established by finance, but the transactional data flow from the rest of the enterprise.

Why weren't all these systems integrated from the start? Well financial ledgers represented a defined and established body of knowledge (accounting standards) and the first systems to use computer support. MRP and closed-loop MRP were young emerging concepts that when stable and proven made sense to be electronically integrated with the financial systems.

Improved Support of Purchasing/Procurement

As with customer order management, expanding the capabilities of closed-loop MRP to include purchasing processes made sense. The supporting processes were similar, as was the type of data (Table 6.1).

Slowly but surely there was a trend toward buying more items versus producing items internally, as companies started to focus on their core competence. That is what they did best, and developing enhanced computerized support of the purchasing processes and integrating them with closed-loop MRP just made sense.

Additional Methods of Planning

MRP and closed-loop MRP were designed around the assumption that a company built physical devices in a predetermined lot or batch size and

TABLE 6.1

The Similarities between Customer Data and Transactions and Supplier Data and Transactions

Customer	Supplier
Profile the customer	Profile the supplier
Establish customer pricing strategy	Establish target cost
Respond to requests for quotation	Request quotations
Accept customer orders	Award/place purchase orders
Customer follows up to confirm delivery	Follow up with supplier to confirm delivery
Ship product—hand off data to accounts receivable (financial ledger)	Receive product and inspect—hand off to accounts payable (financial ledger)
Receive money—get paid	Pay money

kitted (pulled together) all the items needed to make each batch prior to the production start, and the batch only moved to the next processing step when the entire amount of product was completed at the current one.

This didn't work well with companies that made large volumes of product on highly engineered and specialized production lines. Daily production was a set target; there was no kitting, and no one wanted to report issuing of materials or product completions. Cumulative targets were the norm that aggregated planned production against various time horizons and recorded actual results. Again, some talented professionals figured out how to reconcile these needs with what was proven to work and came up with a style of operating called repetitive manufacturing. Unfortunately this style doesn't fit the needs of those who make products that need to be lot traced, but demonstrates that if the underlying logic of a business system is sound, it can be often adapted to new demands and purposes.

Two other noteworthy planning approaches emerged—just in time (JIT), now often referred to as Lean manufacturing, and the theory of constraints (TOC)—both of which will be covered in subsequent chapters of this book.

Other Styles of Manufacturing

Good news travels fast. Closed-loop MRP worked, MRP2 was even more exciting. But there were two different styles of manufacturing that with some enhancements could utilize the power of this powerful set of best practices that needed to be accommodated. Both had a few common shared requirements.

One was those producing a chemistry-based product, such as food, nutritional supplements, pharmaceuticals, cosmetics, chemicals, etc. The other was the aerospace and defense (A&D) industry. Both were highly regulated and lot traceability a central requirement. Ironically, there were many similarities but still major differences between the two styles of operating, and once again thought leaders or innovators figured out how to accommodate both. While the A&D configuration of MRP2 isn't covered in this book, the configuration supporting food and nutraceuticals will be examined later in this chapter.

Thinking Global

The need to effectively manage resources is universal within the manufacturing industry. Even before the end of the Cold War, a large group of nations often referred to as the West, which included North America, South America, Western Europe, Japan, and other Pacific Ocean nations, operated as a trading block that was very vigorous. There was significant cross-investment by companies headquartered in one country, with affiliate companies in other nations. Although there were some trade obstacles to true free trade, larger companies had the resources to mitigate them and wanted the benefits of MRP2 implemented within their international operations.

While English had emerged as the *de facto* language for international business communications, the need for translation of MRP2 systems and educational material into other languages was obvious, including managing multiple currencies and units of measure and other national taxation schemes.

These requirements created the momentum to translate into other languages, and finding an automated means of doing so presented a technology challenge that persisted for many years.

There was one characteristic of MRP2 that on one hand made it easier to translate, and on the other hand more difficult. MRP2 was based on new concepts and techniques that didn't previously exist for the most part, and new words had to be invented for them. An important dictionary was developed to provide definitions for these new words. The publication was the *APICS Dictionary*, which over decades has continued to be updated. Many people believe this dictionary created the basis for operations and supply-chain management as a profession because practitioners now had a common professional language that was specific and unambiguous.

A brief recap:

- MRP2 incorporated a system that allowed senior management to directly drive the organizational planning and execution plan. This system is sales and operations planning (S&OP).
- MRP2 had significant enhancements to support both customer order and purchasing/procurement management.
- The integration of financials ledgers with MRP2 created a unified view of financial activity and performance.
- Additional methods of planning were added in recognition that one planning technique didn't fit all. The concept of a hybrid planning approach took root.
- MRP2 was enhanced to support other styles of manufacturing, i.e., recipe or formula-based products such as food and nutraceuticals.

- Corporations and MRP2 software developers recognized that the body of knowledge that MRP2 represented needed to be translated into other languages and support international requirements.

Looking ahead, the rest of this chapter will cover:

- The historical return on investment (ROI) from effective use of MRP2.
- Enterprise Resource Planning (ERP).
- The differences between standard ERP and process batch ERP.
- Is ERP compatible with Lean?
- The emergence of the current supply chain philosophy.
- Familiarity may breed contempt, but ERP is still critical to competitiveness today.
- How to get up to speed quickly on the best practices embedded in ERP.

The ROI

While all five performance measures in both studies are important, improving inventory management to reduce surplus (unwanted and not needed) inventory stood out as the biggest contributor to freeing up cash, and inventory turnover the best indicator of progress. The calculation for determining inventory turnover is: Divide the average inventory level into the annual cost of sales (Figure 6.8).

	Anderson, Schroeder, Tupy & White Study 1981			Clemson Study 1986		
	Pre-MRP Actual	Current Actual	Future Estimate	Pre-MRP Actual	Current Actual	Future Estimate
Inventory turnover	3.2	4.3	5.3	4.5	7.9	11.2
Delivery lead time (days)	71.4	58.9	44.5	55.6	41.7	31.8
% Meeting delivery promises	61.4	76.6	88.7	73.9	88.6	94.6
% Of orders requiring splits—unavailable	32.4	19.4	9.1	29.0	13.5	5.5
Number of expeditors	10.1	6.5	4.6	10.8	5.1	2.1
Sources:						

FIGURE 6.8
Average operating performance—MRP2 users. (From APICS International Conference & Expo, October 21–23, 2007, Denver, CO. With permission.)

The total or real costs of surplus inventory are often overlooked and, when included in calculations, can add up to eighteen percent, in addition to the cost of money. A list of these costs follows:

1. Storage space
2. Item obsolescence
3. Damage to inventory
4. Outside warehousing
5. Personnel accidents
6. Overtime
7. Damage to equipment and property
8. Cost of expedited shipments
9. Extra handling
10. Management time wasted
11. Lost business

Enterprise Resource Planning (ERP)

Clearly MRP2 was delivering significant value to corporations that deployed and managed by it in a disciplined manner. ERP was the next evolutionary step.

In general, ERP evolved more because of newer computer-based information technologies that added more tools to plan, control, and manage the enterprise. What the new computer-based technologies made possible includes:

- Improved integration between the various MRP2 subsystems
- The ability to manage the centralized database more efficiently through relational database management systems (RDBMS)
- The ability to rapidly rewrite software instructions to process data and perform analysis quicker and more efficiently
- Improved human engineering to reflect how those using ERP sequenced through information to make decisions and extensive use of the graphical user interface made famous by Apple Computer and Microsoft's Windows
- Using new emerging tools based on artificial intelligence (AI), such as:
    1. Configurators used during order entry and structuring bills of materials/formulations
    2. Advanced planning systems (APS) that could consider multiple constraints or objectives simultaneously

- Distributing planning and control systems over computer networks, both internal and external
- Information portals that aggregated both structured (MRP2 outputs) and unstructured data from various sources
- Using the web as a communications conduit
- Workflow views of producing a product or service versus the traditional hierarchical views
- Confidence that new emerging concepts and approaches could be accommodated:
    1. The virtual enterprise
    2. Time-based competition
    3. Mass customization

As the reader works through this book, he or she can be assured that ERP continues to evolve to support new competitive challenges.

THE DIFFERENCES BETWEEN STANDARD ERP AND PROCESS BATCH ERP

MRP and closed-loop MRP were designed to plan and control the manufacture and assembly of figurative little red wagons, a physical device such as a toaster, combustion engine, electronic assembly, etc. With the advent of MRP2, products characterized as chemistry or nutritionally based started to be accommodated, such as food, nutraceuticals, pharmaceuticals, cosmetics, and specialty chemicals. A simplistic and limited comparison of traditional (discrete) versus recipe/formula-based products is shown in Table 6.2.

One major difference is in the engineering documentation of an assembled item versus a recipe or formulation-based item. The assembled item starts with an engineering drawing, and the parts list or BOM is derived from the engineering drawing, as is the routing or sequence of operational steps to make the product (Figure 6.9).

A formulation is more interactive in its development because the food scientist/formulator achieves chemistry or nutritional values by manipulating the ingredients used or the amount of a specific ingredient. An example is manipulating the ingredients and their amounts to get the right value for fat content or protein or sugar. One other difference is that the

TABLE 6.2

Comparing Traditional Hard Goods Products with Recipe/Formula-Based Products

Traditional (Discrete)	Recipe/Formula Based
Dimensions	Chemistry/nutritional values
One-time compliance (certification)	Continuing proof of compliance
Use	Apply or ingest or inject
One or two units of measure	Many units of measure possible
Long shelf life (product dependent)	Short shelf life (product dependent)
Bill of material (BOM)	Recipe/formulation and BOM
Traceability—industry/product specific	All products traceable, one up, one down
Fabricate or assemble	Mix or blend
Make a single product	Possible co-products or by-products
Limited regulatory oversight	Extensive regulatory oversight—cGMP, HACCP
Standardized components	For some ingredients variable potency requires blending
Subassemblies	Intermediates
Simple product description and labeling	Comprehensive ingredient and nutritional values labeling
Compliance with labor, safety, and environment laws	Similar to traditional plus allergen management

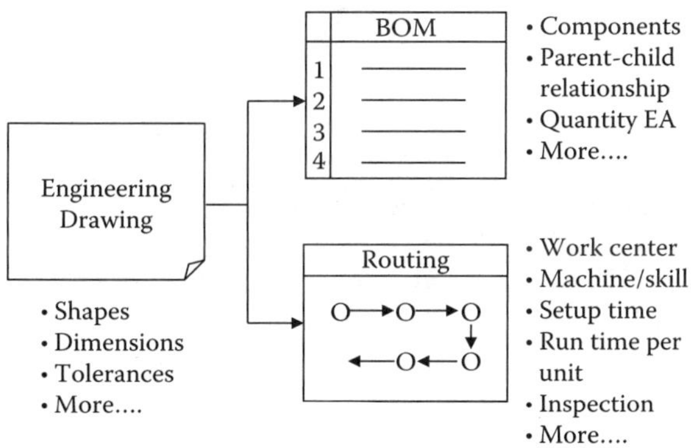

FIGURE 6.9
Product documentation development for a physical device.

recipe or formulation and processing instructions are combined with the recipe because how the ingredients are processed can change the desired outcome in either taste or nutritional values. What the formulator creates is an integrated "whole earth" view of the end product (Figure 6.10).

An example of the integrated nature of a batch process ERP system is in Figure 6.11, which highlights the interactive relationship between the ingredients, the nutritional values resulting from them in the batch, and total raw material costs (Figure 6.11).

Another difference between traditional ERP and process batch ERP is the precision required in the unit of measure in the recipe or formulation

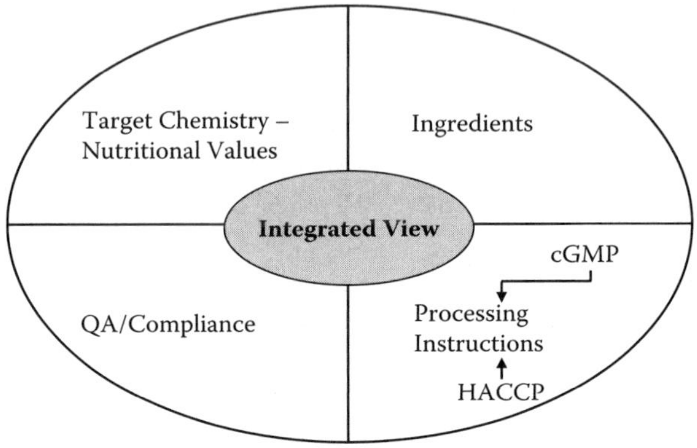

FIGURE 6.10
Product documentation development for a chemistry/formula/recipe-based product.

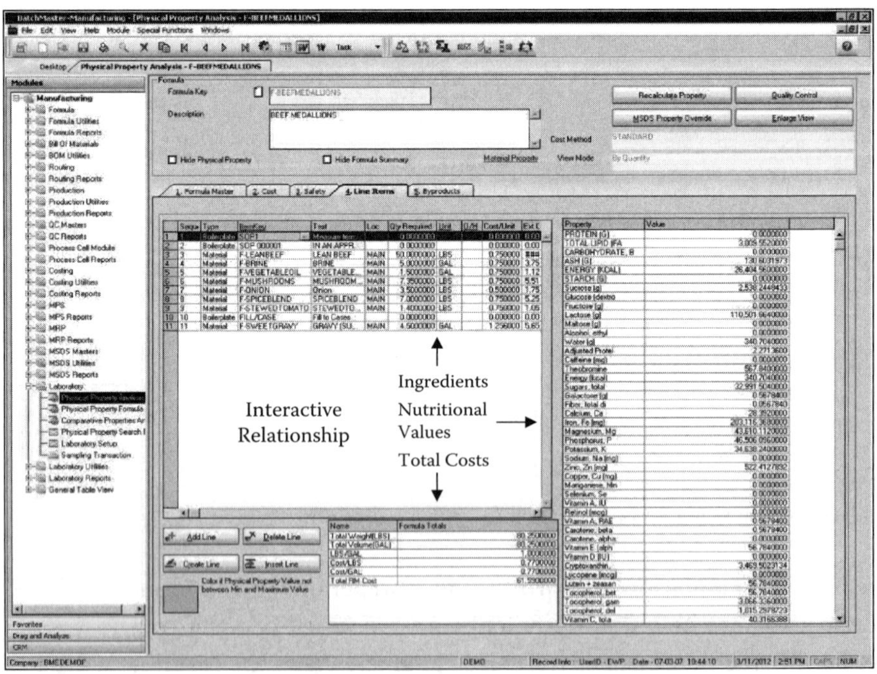

FIGURE 6.11
Interactively connecting ingredients, nutritional values, and costs. (Software images provided by BatchMaster Software, Inc. With permission.)

TABLE 6.3

Information Typically in a Bill of Material (BOM) for a Hard Goods Product

Part Number	Description	Unit of Measure	Quantity per Assembly
123456	Wheel, 16"	Each	4
987654	Bolts, wheel	Each	24
234567	Handle	Each	1
876543	Tires, 16"	Each	4

for ingredients or raw materials. If a little red wagon is being built the parts (ingredients) in the bill of material (BOM) are whole numbers. See Table 6.3 for an example.

In a recipe there is the need for decimal precision that can vary based on what is being made, ranging from two to four decimals (X.XX, X.XXXX) or more. This is necessary to ensure, among other considerations, that the nutritional declarations are complied with for the product.

Every ingredient or raw material in the ERP system's database is unique. The uniqueness is the result of all the item's attributes, such as:

- Description
- Make or buy
- Unit of measure
- Cost
- Revision level
- Planning method
- Costing method
- Minimum order quantity
- Packaging quantity
- Stocking location
- Planner/buyer
- Documents
- Lot traced—yes/no

There are many other data elements that in combination make an item unique, and they are stored and maintained in an ERP data file called the item master. Most are entered when creating the electronic file, and some are automatically calculated by ERP software. The majority of data required are the same for both a physical item and an ingredient used in a recipe/formula, but there are some differences. A few examples:

- Is this item an allergen or does it contain an allergen?
- If so what type?
- Is this item under HACCP controls?
- Does it require lab testing?
- Does it require refrigerated storage?

Another example of the differences between standard ERP and process batch ERP is the tracking of allergens. If an ingredient is an allergen, the item masters for the intermediates or final recipes it's used in are automatically updated to reflect the presence of the allergen; in the jargon of ERP, "it's rolled up" (Figure 6.12).

Working with food ingredients that are defined as allergens is critically important in the food industry because of the health impact on their workers and consumers who have specific food-related allergies. The U.S. FDA recognizes eight foods as being common for allergic reactions in a large segment of the allergen-sensitive population. These include peanuts, tree nuts, eggs, milk, shellfish, fish, wheat and its derivatives, soy and its derivatives, as well as sulfites. All packaged food must have a declaration on it if it contains an allergen and what type of allergen. Even if the food product doesn't contain an allergen, but the facility it was made in does

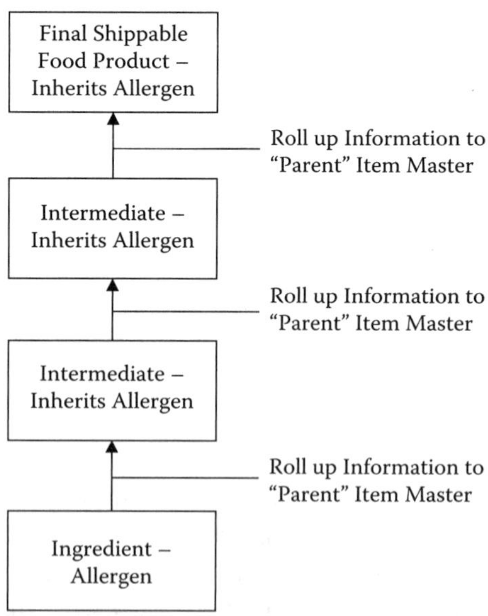

FIGURE 6.12
An allergen ingredient is passed on to its parent, making the parent allergenic.

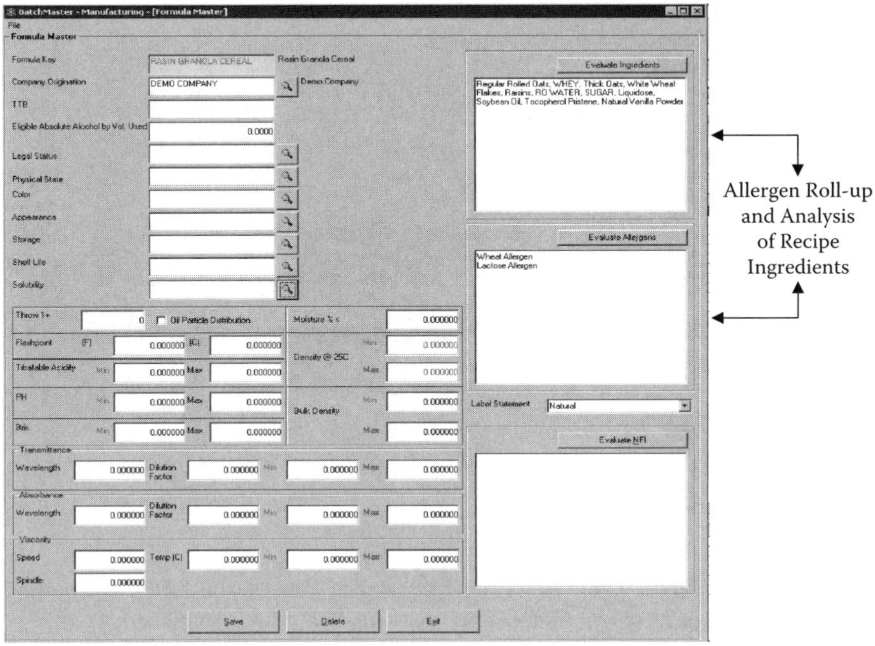

FIGURE 6.13
An ERP system with allergen roll-up capability. (Software images provided by BatchMaster Software, Inc. With permission.)

process other food products with allergens, this also must be declared on the packaging label. Figure 6.13 is a computer system example of the allergen roll-up.

Scheduling in a plant where there is a mix of allergens and non-allergens used in batches requires advanced visibility of the allergen batches, and the need for special handling and post-batch cleaning and sanitation of vessels, and equipment. This requirement is similar to the production environment that manufactures both kosher and non-kosher products that are heated in stainless vessels. When a non-kosher batch is made the equipment must be "kosherized" before a kosher batch can be processed, and that can significantly impact capacity availability. Figure 6.14 is a visual tool to manage this set of requirements.

The important takeaway from this discussion is that process batch ERP has some important capabilities that are important to a company making regulated recipe/formulated products. Trying to use the traditional ERP design instead of process batch ERP will create information gaps that often lead to spreadsheets or "islands of information" that undermine the concept of an integrated enterprise and are an invitation for data quality issues.

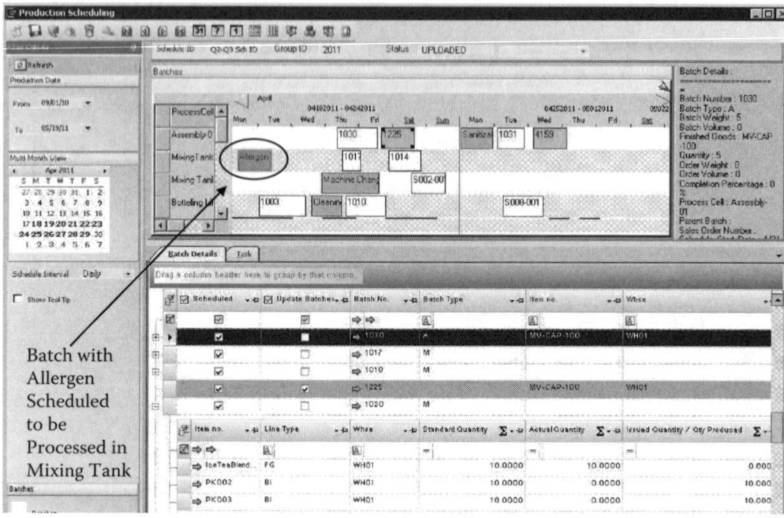

FIGURE 6.14
Visual computerized tools to support allergen management in production scheduling. (Software images provided by BatchMaster Software, Inc. With permission.)

IS ERP COMPATIBLE WITH LEAN?

Absolutely! Lean is focused on continuous improvement and the elimination of waste using proven techniques such as A3, value stream mapping, kaizen, 6S, single-minute exchange of die (SMED), etc. What these Lean techniques need is some level of consistency and predictability in a process. Process batch ERP provides a framework and the consistency needed by these Lean tools. Process batch ERP doesn't know if a process is efficient or inefficient. It just manages the process as designed, which in all likelihood has opportunities for improvement. Those opportunities for improvement are the focus of the Lean toolset of techniques:

- A3 is about framing a problem, immersion in the details, questioning assumptions, and developing strategies and tactics to resolve the problem.
- Value stream mapping is centered on documenting a process, including the value-added time and the idle time (waste), which highlights the parts of a process that need to be re-engineered.

- Kaizen is the re-engineering of activity and is historically unique in that the people who perform the work are those who improve the process redesign to remove waste.
- 6S is not only about uncluttering a workspace but uncluttering the mind.

After a process is improved, process batch ERP takes over managing the process by measuring lead time and cost variances; they function as alarms to indicate any deterioration in process performance. Even if a process maintains its improvements it is still subject to the Pareto principle (80/20 rule) and will in all probability be analyzed, dissected and re-engineered again to reduce waste; that's why it's called continuous improvement.

There is one inventory replenishment technique synonymous with Lean called kanban. Kanban is designed to pull or start replenishment activities based on actual demand, in contrast to MRP, which calculates in advance the material requirements for ingredients, intermediates, and packaging materials and broadcasts a suggested plan to the enterprise. MRP is often referred to as a push technique.

An item being managed under kanban control uses a physical card attached to a predetermined amount of material. When the material is pulled to satisfy a production order, the card is sent to the appropriate work cell, department, or supplier (via purchasing) as authorization to replenish the amount stated on the kanban. Kanban started out as a manual system, and most recently many organizations have utilized it in electronic form with the use of bar coding of the kanban card, and when the card is scanned an electronic signal is sent to the internal or external supplier.

The goal of kanban is to:

- Reduce inventory investment by leveraging the increased velocity or throughput of product through the entire manufacturing and procurement process made possible by kaizen events. A successful kaizen event will reduce wasted time and shorten lead time. Lead time and batch size are linked; the longer the lead time, the larger the batch. A reduced lead time can result in smaller replenishment batches.

  One potential obstacle to leveraging the kaizen improvements is the setup or changeover time to make a batch of product. Setup time is expensive and adds cost to each unit of product processed. A larger batch spreads the setup costs and reduces the amount of cost per unit. The good news—there is another Lean technique designed to reduce setup times: single-minute exchange of die (SMED).

The goal of SMED is a one-minute setup time. In practice difficult to achieve, but with every minute shaved off the setup time for a batch, the economics for a smaller batch becomes acceptable.

- Tie inventory consumption to actual demand.
- Store as much inventory at the point of use as possible.
- Create a visual status on replenishment needs and activities.

While kanban is widely used throughout many different industries there are two levels of usage; one is simplistic, and the other more sophisticated and intensely focused on inventory reduction.

The simplistic use is the concept of two-bin inventory storage. Each bin has a kanban card and has enough material in it to match estimated consumption over lead time to replenish. When the contents of one bin are consumed, the replenishment signal is sent. Generally this approach is used on low-cost commodity items.

A more sophisticated approach is the use of more than two cards, and it's commonly called the multicard approach. The goal is to continuously reduce inventory and shorten replenishment cycles. Organizations that have developed a deep Lean culture and have years of experience in eliminating waste have the expertise to use this approach and have developed an intricate network of internal and external suppliers in constant communication with each other, often using electronic workflow communications to keep things simple. It is not unusual to find that their warehouses and stockroom have a small physical footprint, and they believe that handling inventory adds no value, which is why the preference is for storing inventory at the point of use. Please note that companies that are obligated to keep lot traceability records have successfully used kanban.

But is kanban compatible with ERP? Yes, it's another tool that can be used to manage inventory replenishment activities. Many companies will use a mix of MRP and kanban to manage, and most, if not all, established ERP software systems support kanban.

Lean and its contribution to not only operational performance but also food safety and sustainability will be examined in more detail later in this book.

THE EXTENDED SUPPLY CHAIN AND THE CORE COMPETENCE CONCEPT

Up until approximately twenty years ago many, if not most, organizations strived for enterprise integration *and* control over all productive assets and capabilities needed to produce a final product. A classic example is the Ford Motor Car Company, which owned assets starting with the mines providing iron ore to make steel, progressing to rolling a finished, ready-to-be-sold car off an assembly line. Think field-to-fork managed by one corporation in the food industry. Yes, companies have always purchased items from specialists we now call suppliers, but the emphasis was on internal self-reliance. Despite the development of ERP, the task of managing all this complexity proved daunting. ERP was created to deal with the activities within the four walls of a factory and had a figurative vertical perspective, which is a planning process that went through progressively more levels of analysis, starting with product families and ending with detailed recommendations that could be executed by the workforce.

In the late 1980s, Mr. Charles Handy forecasted that the future enterprise would compete on the basis of its core competence and out-source anything that distracted from that focus. To illustrate what it would look like, he used the analogy of a three-leaf shamrock to illustrate the future organization:

- The first leaf was the core team of highly focused individuals representing the organization's core competence. They were full-time trusted employees critical to the success of the enterprise.
- The second leaf was specialists producing or providing services needed by the organization but not centered on the core competence or basis of competing of the organization. They were now subcontractors and in a prior era would have been employees of the enterprise.
- The third and final leaf was the skilled workers brought on board as needed but not assured a permanent position.

This was an appealing operating model because it stripped away complexity and pushed accountability to others. It was interesting to see that many companies wanting to adopt this new way of thinking had trouble identifying their core competence or value proposition, but with time they planted a stake and slowly out-sourced whatever wasn't within their specialization. What is truly amazing is the speed in which this operating philosophy took hold, and today the concept of a virtual organization is embedded in our thinking.

Rapid Globalization

Prior to the end of the Cold War the world was divided into three distinct camps:

1. The figurative Western world and other pro-capitalist countries
2. The Soviet Union and the People's Republic of China and the countries affiliated with or simpatico to their ideology
3. Countries that proclaimed neutrality in the competition between these two major philosophies

In the end rationality and economics won and the Cold War ended. Suddenly the number of potential customers, suppliers, and collaborators increased dramatically. The increasing acceptance of the core competence model, coupled with the myriad choices of subcontractor, made for a disruptive confluence.

The Internet

Technology can be very disruptive and change the competitive landscape. Such was the case with the Internet, a "gift" from the U.S. government. Once used for research collaboration and military purposes, it was literally donated to the commercial community. Within less than a decade it became a global communications backbone that became secure, immune to time zone differences and other obstacles. It's also inexpensive, which always accelerates acceptance of anything useful.

Combine these three change agents and manufacturers had strong motivation to migrate to a new supply chain philosophy.

ERP was and still is important because every organization participating in a supply chain has to perform at a high level without fail. A chain is only as strong as its weakest link.

The problem was, and still is, we've now developed complex horizontally oriented supply relationships without the clarity or visibility we have with ERP. Poor visibility across supply chains is still a major problem cited by respected industry analysts focused on supply-chain management. Using the development and evolution of ERP as a reference point, current supply

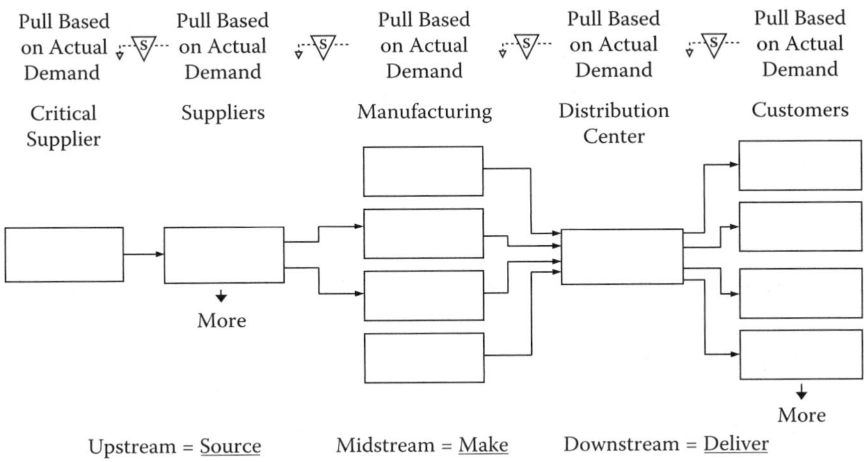

FIGURE 6.15
Demand-driven integrated supply chain.

chain management systems are close to the closed-loop MRP era in terms of evolution, and there are best practices currently being developed that will resolve many current supply-chain management challenges. Two of the most pressing problems are lack of visibility by upstream suppliers more than one level up and risk management (Figure 6.15).

There are at least four opportunities related to improving supply-chain management:

- The SCOR Framework developed by the Supply Chain Council, a nonprofit professional association.
- The APICS Certified Supply Chain Professional (CSCP) certification that is supported by courseware taught by qualified instructors in "live" classes at approximately two hundred different locations in the United States alone, as well as in other countries. There is even an Internet-based option (distance learning) for those who prefer taking the preparatory courses at a flexible pace and schedule. The CSCP course is centered on proven best practices with an emphasis on the integrated supply chain and managing all aspects of it.
- The APICS Certified in Production and Inventory Management (CPIM) certification. Each factory or distribution center is a figurative link in the supply chain and must be managed in a disciplined fashion with proven best practices. CPIM examines in depth all the philosophies, methodologies, and techniques proven to produce

positive bottom-line results, including how to choose the appropriate technique for a given situation and the prerequisites for success. Accessing CPIM educational courses is similar to that of accessing CSCP courses. All the processes inherent in process batch ERP are based on the concepts and techniques incorporated into the CPIM curriculum.
- Universities that offer a certificate program on operations or supply-chain management. The goal of these educational offerings is for the student to gain a working knowledge of the principles and fundamentals not necessarily for certification, but for providing a solid foundation for pursuing that later. Often these programs are cosponsored by the university and APICS or other professional associations. Some universities offer advanced certificates.

Beyond these four opportunities are other possibilities, such as the Institute of Supply Management (ISM), which focuses on but is not limited to procurement sourcing and the contract (legal) side of buying. Others include the Council of Supply Chain Management Professionals (CSCMP), whose roots are in logistics; the Manufacturing Execution Systems Association (MESA), which specializes in factory automation; and the Association for Manufacturing Excellence (AME), which promotes continuous improvement and Lean thinking.

A resource often overlooked is the American Society for Quality (ASQ). A Lean efficient supply chain with the excessive inventory removed from it requires quality products flowing through it. Quality just doesn't spontaneously happen; it requires the application of best quality techniques and practices, and ASQ is the acknowledged thought leader.

For those who prefer learning in smaller increments and on a semirandom basis there are many focused magazines that are listed in this book's appendix.

SUMMARY

ERP evolved over a fifty-year period, starting as a system (MRP) to calculate detailed time-phased inventory requirements based on actual customer orders or on forecasts against existing available inventory both in the stockroom and on order. MRP created a suggested plan for every item required, which included the amount, order start date, and order finish date. The catalyst for inventing MRP was the high cost of excessive inventory.

Earlier pioneering corporations using MRP made substantial operating improvements, including inventory reductions and on-time delivery to customers. The carrying costs for surplus inventory have historically been and continue to be very expensive. Success and experience using this new tool fostered innovation and the successor to MRP was closed-loop MRP.

With more success, innovators once again added capabilities to closed-loop MRP to support additional departments and processes. This necessitated a new name to define this latest interaction of a planning and control system, and MRP2 became the new baseline for innovation.

During the MRP2 phase of the evolution to ERP, innovators and leaders in best business practices thinking started to adapt MRP2 to the requirements of different types of manufacturing, i.e., food and beverage, aerospace and defense, pharmaceuticals, as well as the needs of distributors.

Newer information technology was the final stage of evolution to ERP. It facilitated faster development of computer software and centralizing of data for the enterprise and a more intuitive user interface influenced by the Apple Macintosh and MS Windows graphical interface. At this point, compared to all other sectors in the economy, the manufacturing industry had the only comprehensive and integrated planning and control framework (ERP).

As ERP matured and as the pool (market) of large and medium-sized companies buying and implementing ERP became saturated, the price of ERP software and the related equipment to maintain it declined and ERP became affordable and a compelling investment for smaller companies.

Process batch ERP is a special adaptation for companies who make highly regulated recipe or formula-based products. ERP's need for data accuracy, integrated data, defined repeatable processes, and discipline in managing processes to achieve world-class operating performance is in sync with food (and nutritional supplement) safety regulatory requirements. Process batch ERP is an opportunity for those manufacturers of recipe/formula-based products who are still using multiple disconnected electronic spreadsheets to manage their business in order to improve business performance and lower food safety risk as well as the cost of compliance.

Getting a workforce up to speed on the best practices embedded in process batch ERP is straightforward. There are multiple cost-effective sources of education and training: professional associations, universities and colleges, for-profit education providers, independent consultants, and government-sponsored low-cost consultancies.

The benefits of applying Lean manufacturing techniques are well established. In fact, they are also useful in streamlining administrative processes and the recipe/formulation development processes. Such techniques are:

- A3
- Value stream mapping
- Kaizen events
- 6S
- Visual workplace
- Kanban

These techniques are best leveraged in a company committed to a culture of continuous improvement.

Lean and process batch ERP are compatible and synergistic. Process batch ERP manages and monitors processes and the data associated with them. It provides a stable framework. Lean improves processes striving for more efficiency, improved quality, and increased velocity of work and workforce empowerment. Once a process is improved, process batch ERP takes over and monitors it once again. From a management perspective Lean is appealing in that the techniques are not complicated, they work, and their effectiveness increases with use.

QUIZ

6.1. Name three practices emphasized in ERP that are shared by best practices in food safety.

6.2. ERP is a framework for organizing, defining, and standardizing the business processes necessary to effectively plan and control an organization so the organization can use its internal knowledge to seek external advantage.

☐ True ☐ False

6.3. ERP is the result of over fifty years of refinement and evolution.
☐ True ☐ False

6.4. In the late 1940s pent-up demand for goods and the lack of systems to balance demand and supply forced organizations to figuratively surround themselves with costly inventory, and in response, the search for better inventory planning methods started.
☐ True ☐ False

6.5. Material requirements planning (MRP) was the first step in ERP's evolution and was designed to calculate down through a multilevel recipe (BOM) the inventory that was needed to support actual customer demand.
☐ True ☐ False

6.6. What four root causes of success or failure in using MRP by early adopters were uncovered in an in-depth analysis?

6.7. What two systems were added to MRP to increase its successful use?

6.8. What name tag was put on the next evolutionary step that enhanced MRP?

6.9. With widespread success closed-loop MRP was enhanced and renamed MRP2. List the seven enhancements it includes.

6.10. Many studies on the benefits of MRP2 were conducted; one is referenced in this chapter. What number of inventory turns improvement was noted in the report?

6.11. What is the cost of excessive inventory, inventory beyond immediate needs (perfect demand-supply balancing)?

6.12. MRP2 evolved into ERP, which is the current *de facto* framework to manage a manufacturing enterprise. Most of this evolution was the result of leveraging advanced information technology. Name three out of the nine contributors to the evolution of ERP mentioned in this chapter.

6.13. Name four of the many different processes or requirements that process batch ERP supports that traditional hard goods ERP doesn't.

6.14. ERP and the Lean management system are complementary.
☐ True ☐ False

6.15. What concept foretold the extensive out-sourcing that occurred during the last decade?

6.16. What three developments have led to the rapid development of extended and complex global supply chains?

6.17. The move to extended supply chains hasn't been problem-free; besides risk management, what other opportunity for improvement has surfaced?

6.18. With the move to extensive out-sourcing and the extended supply chain model the importance of ERP diminished.
☐ True ☐ False

6.19. Name the four major channels or forums to improve supply-chain management.

The answers to these questions can be found in the Appendix along with definitions, website addresses, and recommended reading.

7

How to Organize the Workplace: Lean Thinking

Lean thinking and *continuous improvement* (CI) are often used as interchangeable terms, but for the experienced Lean practitioner they are separate although interconnected. Lean is best described as a philosophy, an operating system, and a management system, and CI is a toolkit of improvement techniques used within the Lean management system. Companies who have committed to the Lean management system and have experienced tangible results will refer to this commitment as the Lean journey, and it is a critical part of their corporate culture and a distinct thread that runs through it. This is the difference between doing Lean as a project, which is what some organizations unfortunately have done, and Lean as a management system (Figure 7.1).

At the core of Lean is respect of the individual and the obligation to enable each person to solve problems and improve the work he or she does and ultimately be self-managed, not via the traditional command and control style. What the various CI tools do is help individuals improve critical thinking, identify and understand value from a customer's perspective, understand processes, and identify opportunities to eliminate waste in the form of unnecessary:

- Overproduction
- Waiting
- Inventory
- Motion
- Transportation
- Defects
- Extra processing
- Underutilized people

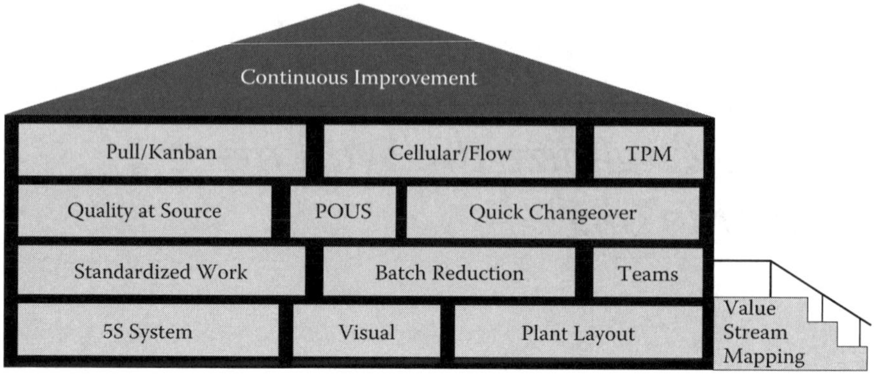

FIGURE 7.1
The House of Lean. (From Ron Calame, California Manufacturing Technology Consulting® (CMTC), Torrance, CA. With permission.)

With the opportunities for improvement identified there is an approach (tool) to reducing the waste on a selective basis. When the improvement is designed and implemented, so are mechanisms to maintain the improvement.

While these CI techniques are often thought of as factory floor-focused, they have proven to improve all processes, such as product and service design as well as all administrative processes. There is one methodology that even helps a supplier accurately determine what customers really want, and it is a real catalyst to innovation.

Many of the Lean techniques are often associated with Toyota Motor Company and Japan, but there were numerous companies in North America that developed techniques similar to and without any knowledge of Toyota's operating model starting in the late 1960s. Lean is really based on common sense, but unfortunately the statement "common sense is an uncommon quality" rings true. The benefit from publicizing the Toyota approach to running an enterprise was that it laid out for examination commonsense techniques for the frugal use of resources and the underlying reasons for using each of them.

One appealing aspect of Lean is that it is manufacturing industry neutral. It works for companies that build airplanes, toasters, computers, and those that make food products.

Lean has applicability and has proven to work in many sectors outside of manufacturing industry, industries such as public services, educational services, healthcare, service industries, etc. It delivers substantial value in almost any situation where one or more individuals organize to perform work and want to eliminate waste, even hidden waste, waste that is not obvious and figuratively "staring at you."

Another positive attraction is that Lean isn't computer based and doesn't require huge financial investments. It does require patience since a workforce is being progressively retooled to think and work differently and Lean is geared toward continuous improvement, which means the status quo is always being challenged.

While the majority of this chapter is centered on objectively analyzing and organizing the workplace, some of the conversation is directed toward a brief revisiting of process batch Enterprise Resource Planning (ERP) because it is a stable framework for Lean/CI to operate in.

As each CI technique is examined, it is looked at from two perspectives: the practitioner perspective and the executive view. Generally speaking, the practitioner perspective follows the why, what, and how convention. Executive engagement is centered on defining strategic direction, designing the management system, and providing necessary resources. In a Lean culture, leadership from the executive suite means that the board of directors (BOD), chief executive officer (CEO), and management staff demonstrate leadership by:

- Acquiring a working understanding of the Lean management system and its CI techniques, not necessarily at an expert level, but each understands the purpose, the essence of the CI tools, and the obligation to enable each associate worker in the enterprise.
- Showing a visible and obvious interest in what's going on in the workplace, not through discussions in the office, but by personal observation and interaction with the workforce. Some might call this walk-around management, but it's more than that.
- Investing in workforce education. Lean is best learned not in the traditional academic framework, but through a combination of simple conceptual education, doing, and mentoring. It's inexpensive to administer, has an excellent return on investment (ROI), and doesn't have a shelf life expiration date.
- Establishing performance measures that are consistent with and fully support the Lean journey is key to ongoing sustainment and full realization of improvements.

The topic of Lean leadership is very important and beyond the scope of this book, as are the entire Lean management system and all the supporting CI techniques. The intention of the author is a brief introduction to Lean and CI. Going beyond this introduction, there are many sources of information and respected texts listed in the Appendix that are recommended for future reading. A general flow for this chapter is shown in Table 7.1.

TABLE 7.1

The Purpose and Use of the Techniques and Tools to Support the Lean Management System

Technique	Purpose and Usage
Techniques That Standardize Problem Solving and Improve Critical Thinking	
PDCA (plan-do-check-act)	Powerful integrated general improvement or problem-solving technique. The cornerstone for continuous improvement.
A3	A technique that supports PDCA and is often used by seasoned Lean veterans to coach and mentor others and increase their critical thinking skills.
Techniques to Understand What Customers Truly Value	
Voice of the customer (VOC)	Understand in a systematic manner the attributes a customer values most in the type of products and services the organization provides. Quality functional deployment (QFD) and its most recent "offspring," outcome-driven innovation (ODI), are excellent and proven methodologies.
A Technique to Analyze and Visually Depict an End-to-End Process Flow (Value Stream) to Identify Opportunities for Improvement	
Value stream mapping (VSM)	Provide a factual picture of the process flows that support delivery of customer value and the cash-to-cash cycle. Identify and prioritize opportunities for improvement. VSM has two views: current state and future (desired planned) state.
A Team-Oriented Technique to Significantly Remove Waste in a Work Area or Process Step	
Kaizen events	A team-based methodology to act on the opportunities for elimination of waste identified in the current-state VSM or A3 to achieve the targeted future state. An event has specific and significant measureable improvement targets established in advance of the event. A kaizen event is also used to correct or minimize a safety or ergonomic issue in support of the sixth S (safety) in 6S. A kaizen event in a work area is generally not a one-time event but conducted repeatedly as needed over the course of time. For food, beverage, and nutraceutical manufacturers, a food safety practitioner should be a member of any kaizen event that addresses quality, manufacturing, or recipe/formulation management processes.

(continued)

TABLE 7.1 (cont.)

The Purpose and Use of the Techniques and Tools to Support the Lean Management System

Technique	Purpose and Usage
Kaizen opportunity	A kaizen opportunity is an improvement that generally is identified, tested, and implemented by one individual or a small group within a work area. Often the improvements are small, but with others cumulatively related improvements create significant value. They are spontaneous in contrast to a structured kaizen event. Those experienced with kaizen are skilled at offline testing and experimentation without impacting current methods of performing work. Some of these skills are acquired through participation in formal team-oriented kaizen events. Because of food safety considerations it might be prudent to have all suggested changes quickly reviewed by internal food safety practitioners.
Continuous Improvement (CI) Techniques Used to Implement Change Directed Toward Waste Elimination	
• 6S	These are the tools available to a kaizen event team to support continuous improvement.
• Workplace organization • Teams • Manufacturing cell • Visual workplace	Therein lies the proverbial what came first dilemma, the chicken or the egg? Each of these techniques has merit even without a kaizen event, as they are a focused collection of best practices. If they are in place within a work area, even if in less than ideal form, the implementation of the improvements surfaced by the kaizen event will be, in all probability, shorter than if not.
• Standard work • Quick changeover (SMED)	If these techniques haven't already been deployed, one or more of these are still the best tools to implement waste reduction as surfaced in a kaizen event.
The Compatibility of a Computerized Data Repository and Planning System with the Lean System	
Process batch ERP	A brief discussion on the compatibility of Lean and CI with process batch ERP.

PDCA (PLAN-DO-CHECK-ACT)

Technique Overview

PDCA is synonymous with Edwards Deming, the thought leader on quality concepts. He greatly influenced the Japanese and most of the rest of the manufacturing world. Deming's PDCA model was an enhancement of the work done by a gentleman named Shewhart. PDCA is based on the scientific methodology that is noted for its simplicity and often referred to as an improvement engine. It is a disciplined, iterative four-step management process with successive cycles.

Each cycle improves the knowledge about the system being targeted for improvement and drives the refining of the goal(s) because of the recently acquired in-depth knowledge of the system targeted for improvement. PDCA is the genesis of the continuous improvement philosophy. The attractiveness of PDCA is that it works not only for any type of manufacturing environment, including the manufacture of recipe/formula-based products, but also for any business environment or industry. There is a predecessor step to plan in the methodology, that is, the recognition that an opportunity to improve exists in a specific area. The four iterative PDCA steps:

- Plan
  1. Define the expected output or goal of the PDCA cycle.
  2. Establish the objectives and processes needed to achieve them, including the data that must be measured.
  3. Review the plan for completeness and accuracy.
- Do
  1. Conduct a small-scale implementation of the plan.
  2. From it collect data for analysis.
- Check (study)
  1. Measure results and compare against expected results.
- Act
  1. Analyze differences between the actual results and expected results.
  2. Determine the causes for the differences.
  3. Refine the plan and execute the cycle again until the targeted goal is reached (Figure 7.2).

PDCA is a basic approach that must be mastered to lay a foundation for other Lean techniques and tools for creating a Lean workplace and organization, an organization that is competitive, profit making, and disciplined and effective in its food safety regulatory compliance efforts.

There is another technique that is designed to help frame an objective or problem, clearly state it, force or encourage objective, unbiased research, challenge premature conclusions, create a visual plan that is easy to comprehend, and secure buy-in from those affected by the proposed plan. This technique is called A3—a strange name, but great technique.

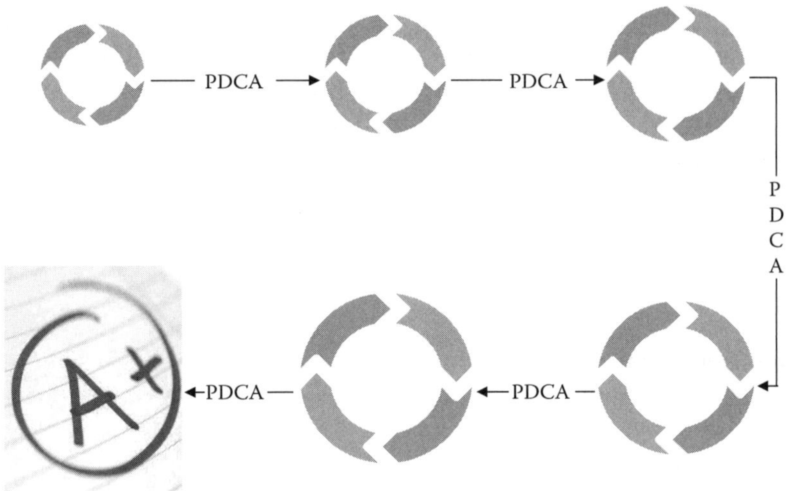

FIGURE 7.2
The PDCA iterative cycle.

Management Considerations

As with all techniques used within the Lean management system, there is generally little capital equipment investment required. Education is required for all Lean techniques, in some cases conceptual, and for many others, in the form of coaching by those designated as internal teachers/mentors. The good news is that all CI techniques are straightforward and readily grasped. There are numerous cost-effective educational sources on all of the Lean/CI techniques, including professional associations, universities, industry trade associations, government-funded organizations, for-profit educators, and magazines such as *Industry Week*.

When a company starts to use any CI technique it is often beneficial to bring in an external consultant to help accelerate practical usage of the technique. Generally a consultant's involvement is for a short period of time.

Skill in the application of any CI technique is acquired by doing, not passive learning.

Since PDCA and A3 are foundational tools within the Lean management system, the first "students" of PDCA and A3 should be the management team, ranging from the CEO through to those with leadership responsibility. This will send a clear signal to the organization of management's commitment in the Lean journey.

The costs for bringing PDCA into an organization are centered on education and some consulting, and a modest support budget should be developed. The Association for Manufacturing Excellence (AME), which stresses company-to-company sharing of the dos and don'ts associated with the ongoing implementation of the Lean management system, is an excellent source of information about what others have budgeted and their achievements.

A3

Technique Overview

A3 is a structured methodology for utilizing the PDCA cycle. With use, it instills a new way of thinking and approach to problem solving. It enhances the critical thinking ability and effectiveness of those using it. Because of that, it is an excellent mechanism for those with supervisory responsibilities to coach and mentor team members.

The name A3 is based on the size of paper used to develop the plan. A3 is the international designation for 11 × 17 in. paper. It has a specific format:

1. Theme
2. Background
3. Current condition
4. Target condition/measurable objectives
5. Root cause and gap analysis
6. Countermeasures
7. Plan
8. Follow-up actions

The size of the paper creates a deliberate constraint when one considers all the information required by an A3. This combination of size and information requirements forces concise statements. The information flow in an A3 easily surfaces information disconnects, just as grammar rules help us communicate clearly. A3 also has a visual component, which is important in helping others understand the problem, the supporting information and data, the processes used, and finally the plan. There is an old saying: "A picture is worth a thousand words."

The power of A3 is that it stresses the need to:

1. Not draw premature conclusions
2. Not take convenient shortcuts in gathering information but personally observe activities and talk to those closest to those activities
3. Reconfirm relevant information and observations
4. Think in terms of cause and effect, symptoms and root causes
5. Be concise with words
6. Use graphical representations to reinforce statements
7. Be complete and logical
8. Secure peer buy-in
9. Progressively build consensus
10. Be iterative in developing the A3; it is a methodology supporting the PDCA cycle

A quick overview of each of the "commandments" follows:

1. Don't draw premature conclusions. There are two beliefs that permeate Western society today: "Time is money" and "don't reinvent the wheel." Both have merit but can have the undesirable effect of relying on partial information, which is often secondhand and encourages shortcuts. Also, when someone has had success in addressing problems in the past and is called upon to fix a problem, they often, because of their previous success, tend to be proscriptive and rely on their instincts without doing the information gathering due diligence required for success. Problems or opportunities for improvement that warrant the investment of resources should be approached in an objective and methodical manner.
2. Don't take shortcuts in gathering information. We live in a data-rich world, so rich it's a form of pollution and it overwhelms. Without even thinking about it we want to get it summarized and sanitized as a defensive mechanism. Unfortunately information that is an abstraction is inadequate for understanding the essence or boundaries of a problem or revealing an improvement opportunity. To understand a situation, it has to be experienced firsthand. A good description—"marinate in it."
3. Reconfirm information gathered. It is easy to misunderstand information provided by others. Solving a problem requires operating off an accurate understanding of the situation. A simple paraphrasing

of a statement or discussion on information with those involved in a situation and asking them if what was paraphrased is accurate can be a real eye opener. Think of reconfirming as a quality check and opportunity to get a more granular understanding of the situation.

4. Get to the root cause. In Lean practitioner circles there is the approach, a questioning technique to determine the root cause of a problem. It's called the five whys. A complement to the five whys is fishbone diagramming. Solving a problem requires getting a firm understanding of the root cause(s). That is the purpose of the five whys questioning scheme and fishbone diagramming. Fixing symptoms is a problem-solving dead end. Getting individuals to think in terms of cause-and-effect relationships often takes some training, but is worth the effort.
5. Be concise. Many years ago the author was involved with a group of professionals trying to improve communications within the group. A facilitator was brought in. In one simple exercise conducted by the facilitator it was demonstrated that approximately fifty percent of a verbal communication was misspoken or misunderstood. With some simple training everyone became more aware and focused on clear unambiguous communications and ultimately became a high-performance management team.
6. Use graphical representations. A great analysis and plan is useless unless key individuals understand it. Without their understanding there will be no action. People have a mix of learning modes. Some are inclined to learn through the written word. Others need to hear the communication; they are called auditory learners. Many others learn through graphical representations, which, combined with auditory (storytelling), are the tribal and historical methods of learning for humankind. There are other methods of teaching beyond this, but they are difficult to administer. The important point is that a picture is worth a thousand words, and A3 requires, by design, visuals.
7. Be complete and logical. No one enjoys a book that has chapters missing. An A3 needs to be complete. No one is comfortable with logical disconnects, jumping from one topic to another that isn't logical or transparent. It isn't going to gain support of those who will be asked to implement an A3 plan and maintain it. Going from A to D and not understanding the progression of A-B-C-D isn't going to get supported.

8. Peer buy-in. The coworkers most affected by the problem or opportunity for improvement must agree that it exists. They are closest to it and a valuable source of information. Lean is a team sport.
9. Progressively build consensus. Getting consensus isn't like a light switch that is either on or off. Having a shared view of the problem, the information surrounding it, its impact on conducting business, the cause-and-effect relationship dynamic involved, and the strategy and tactics to mitigate takes time. People need acceptance time to internalize the new information. This doesn't mean that there is a slow tempo to the development of an A3. Once there is consensus the power of the team is focused on solving the problem or seizing the opportunity and doing so with a plan.
10. Be iterative in developing the A3. It is a tool that supports the PDCA cycle and is intended to be refined with successive cycles.
11. Make sure there is an A3 owner *and* coach. Without ownership there is a risk of the opportunity for improvement getting lost in the day-to-day activities, and without a coach the loss of skills growth for the owner.

Most recently the importance of A3 has been rediscovered by companies involved in the Lean journey, and it makes sense to recommend that the reader of this book examine and study it at a detailed level in the future.

A3 is about objectivity, unbiased information gathering, striving to understand cause-and-effect relationships, and concise communications.

Management Considerations

A3's educational and initial guidance needs are similar to those of PDCA. The real investment is a soft one that involves more management time for coaching and mentoring to increase problem-solving capabilities. This is in contrast to the traditional command and control approach that gives only "lip service" to staff development. The cost of using A3 by management is an increased workload, as coaching and mentoring take more time because individuals learn in progressive layers. Developing world-class organizational competence takes patience.

VOICE OF THE CUSTOMER (VOC)—OUTCOME-DRIVEN INNOVATION (ODI)

Technique Overview

Mention voice of the customer (VOC), and many immediately think of market research. It isn't market research in the traditional sense. In fact, voice of the customer has been incorrectly associated with traditional market research. The goal of market research is to determine what consumers will buy. Market research is focused on achieving minor changes in market share within the status quo. VOC is focused on identifying what the current or future customer values, prioritizing the attributes valued, and understanding the acceptable target value for each product or service attribute. It digs much deeper into what the customer perceives as value added. Why the widespread interest in VOC? Because historically eighty percent of all new products launched into the marketplace fail to meet the designer-manufacturer expectations. VOC, when properly utilized, reverses that failure-success rate. It is often referred to as an innovation engine.

Why discuss VOC in the context of Lean or continuous improvement? Two pillars of the Lean philosophy are eliminating waste and delivering more value to the customer from the customer's perspective, not from the designer-manufacturer perspective. When properly deployed, VOC can identify the improvement targets for the Lean enterprise, starting with product development and through the rest of the value chain.

Since VOC is really a framework, the choice of methodology used within it is critical. Traditional market research tools tend to maintain the status quo. In contrast, VOC tools focused on innovation that adds value to the customer and is the only acceptable choice.

There are two distinct and proven methodologies that produce the desired outcome of understanding what the customer values most, and what he or she would be attracted to. Both look not at declarative statements of need from the customer, but probe deeper into undeclared needs. The declared needs are generally understood; they are commodity-like.

Surfacing and satisfying the undeclared needs is the path to adding value to the customer and what separates the innovator from the ordinary run-of-the-mill supplier. The first is quality functional deployment (QFD). QFD as a VOC methodology is mature and extensively documented.

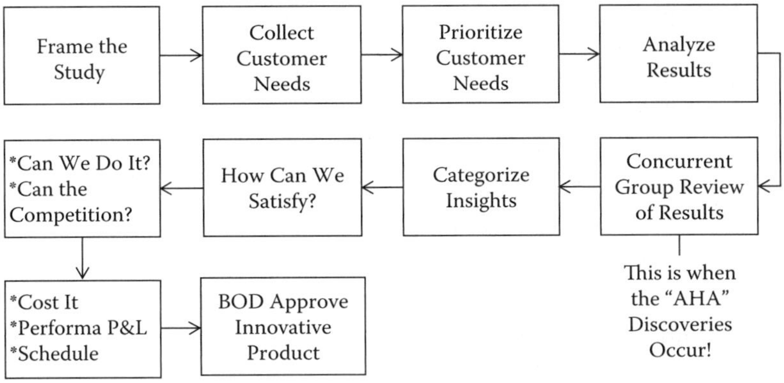

FIGURE 7.3
The VOC process flow.

The other is outcome-driven innovation (ODI) that is, according to its proponents, a refinement of the basic QFD approach. Both stress cross-functional study/research teams, are simple to understand and use, are cost-effective, and with experience, can be used in a pragmatic, flexible manner with minimal computing support. Figure 7.3 is a simplistic depiction of an ODI project.

The U.S. FDA report *Pathway to Global Product Safety and Product Quality* describes the complexity of the current global food supply chain coming into the United States and the challenges this creates relative to food safety. It also addresses survival of U.S. food producers:

> Global changes will have significant implication for (food) manufacturers... there will be increased regulatory scrutiny... intense pressure to *lower costs and improve productivity.*
>
> **—U.S. FDA, *Pathway to Global Product Safety and Product Quality* (a special report)**

There is a "wild card" the U.S. FDA missed. Productivity is a by-product of innovation, not just cost cutting. Innovation through the effective use of VOC differentiates a food or nutraceutical company from its competition, *and* gives direction to its Lean initiatives.

Lean provides efficiency, reliability, and confidence in products and suppliers—innovative products customers value, keeping them excited.

Management Considerations

VOC (QFD or ODI) is project oriented with a distinct process sequence and steps. It requires a cross-functional team approach. Each project (or study) has a goal to identify marketplace innovation opportunities and provide internal prioritization on the application of CI techniques within the enterprise. The ultimate goal is to create products or services compelling to the customer.

As with any new methodology introduced to an organization there are educational and initial guidance costs. The best approach to embedding VOC as a management tool is knowledge transfer from an external expert to key members of the enterprise. Developing self-sufficiency is critically important because the need for innovation and alignment with customer needs is ongoing. Since a VOC project or study can potentially impact the entire enterprise, including product engineering, marketing, manufacturing, quality assurance, supply-chain management, and information technology, it is a significant undertaking. There are soft costs, such as key individuals from multiple disciplines within the company being involved part-time for weeks or, in some projects, months. Travel might also be involved. The good news is that using the methodology gets easier and projects get completed faster with organizational experience.

A major management challenge is to resist making major investment decisions unless a proper VOC study (QFD or ODI) supports them. The hard (real) money costs of one ill-conceived new product launch will be many times more expensive than the soft costs of using VOC.

VALUE STREAM MAPPING

Technique Overview

PDCA and A3 are directed at changing to a new way of thinking that supports continuous improvement, objectivity, clarity of thinking, and organizational buy-in to change. They are improvement engines.

Value stream mapping (VSM) is a technique to identify improvement opportunities in the end-to-end manufacture of product or a set of linked processes that support the enterprise, including administrative and information process flows. Using a saleable product as an example, it starts with the request for shipment by the customer and works upstream (think right to left) through all supporting and manufacturing processes and ends with the receipt of purchased supplies. The important thing is that VSM is trying to visualize the closed loop from request to delivery as well as the issues that affect our ability to effectively and efficiently deliver what the customer is looking for.

A companion tool is spaghetti diagramming, which helps visualize the path a product takes during its "cash-to-cash" life cycle. It is a flexible technique and has other applications or uses beyond that. The ideal path would be linear and move forward. In reality most existing paths or process flows that haven't had Lean techniques applied are convoluted and wasteful with redundant travel throughout an operation. The exception to this would be products flows that are highly automated because there has been a high investment in industrial engineering to protect the capital investment. But in food and supplements manufacturing there are typically many nonautomated paths or flows, particularly with companies that have grown and had to "think on their feet" to solve immediate problems. Figure 7.4 is an example spaghetti diagram.

The purpose of VSM is to provide maximum value to the customer by minimizing waste in the value creation process. VSM is not focused necessarily on cost reduction but on the value experienced from a customer perspective, which might include price, hence cost, but also product quality, reliability of product delivery (dependability), and responsiveness to new requirements and needs. VSM doesn't just look at a single process step, but

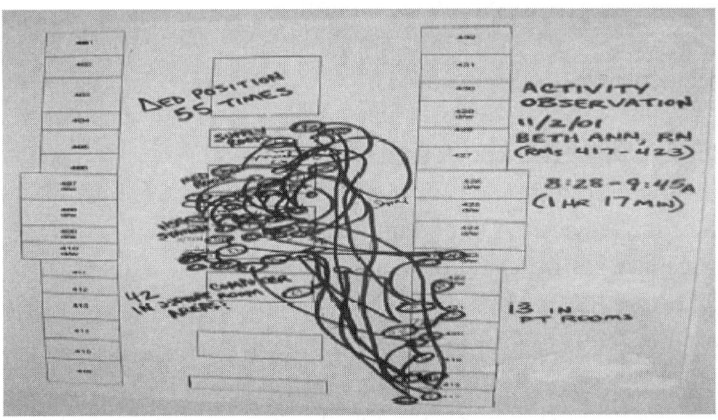

FIGURE 7.4
Spaghetti diagram: before.

all the linked processes that support delivering value to the customer. It is a visualization tool used to surface the issues that impact performance. A VSM can be developed for a single product, product family, performed service, or anything sharing common processes. Think of it as an opportunity for improvement blueprint highlighting critical tolerances.

How Does the Initial VSM Candidate Process Get Selected?

The process starts with management because VSM was designed and intended to be used as a strategic activity in which management is heavily involved. VSM is identifying the *strategic*—what needs to be done; in a Lean environment the front line is empowered to redefine the *tactical*—how the process should run after improvements.

Value stream mapping has two states, the current and the future. The current state must be observed firsthand using the "go see" approach. Secondhand information is not acceptable, even from assumed knowledgeable sources. What the VSM owner must see is:

- Each manufacturing processing step that is required to get the product to the customer with the current post-manufacturing lead times
- Within the end-to-end manufacturing process, each processing step, and for each step the actual key performance and staffing status
- The upstream receipt of raw materials along with the associated lead times
- The forecasting, planning, and control information systems supporting the end-to-end value stream

The basic information required is:

- Number of workers performing the process
- The processing time per unit (runtime or cycle time [C/T])
- The changeover (C/O or setup) time per batch
- Process reliability or uptime
- The buffer inventory
- Other barriers to flow as appropriate (Figure 7.5)

There is a summary line at the bottom of a VSM that has two dimensions; one resembles a high plateau, and the other a low flat valley. The high plateau represents the amount of inventory or time between the process

Pre-Blend
Two (2) People
C/T = 2 Minutes C/O = 2 Hours Uptime = 74% Other Important Defining Data

FIGURE 7.5
Some of the data needed for a value stream map.

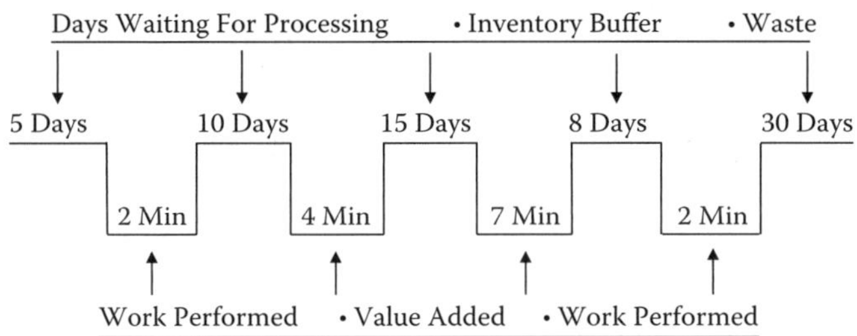

FIGURE 7.6
A combined graphical representation of the plateau (waste) and the valley (actual value added).

steps, which translates into excess inventory or time. The flat valley is the actual processing time (Figure 7.6).

The final layer in a VSM is the information systems supporting the end-to-end process flow:

- Top layer: The information flow.
- Middle layer: Each process step with the resources and capabilities and process time per unit.
- Bottom layer: A visual on the wait time (potential waste) and actual process time per unit.

Figure 7.7 brings all three layers into one cohesive whole. The VSM example used in Figure 7.7 is an indicator of the acceptance of VSM by organizations in many different sectors of an economy. The U.S. EPA, a governmental regulatory agency, doesn't make "widgets" or food, but understands the power of VSM to uncover major opportunities to improving the value chain and eliminating waste.

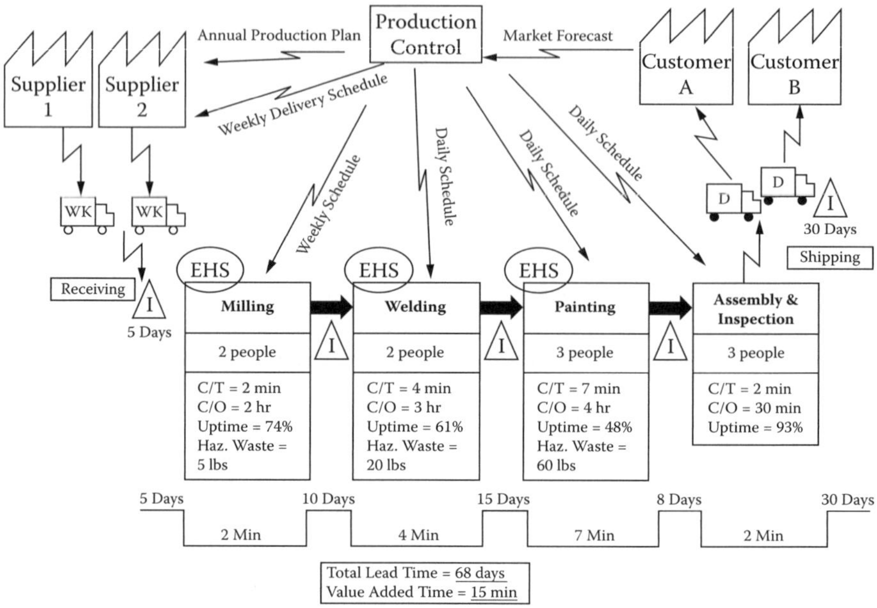

FIGURE 7.7
An example of a value stream map. (From U.S. Environmental Protection Agency (EPA), Lean Environmental Toolkit.)

Although the example used in our discussion is a simple map, VSM is capable of painting a more complex and granular picture if needed. It has its own graphical language, which is presented in Figure 7.8.

Management Considerations

Bringing VSM on board as a CI tool doesn't require much economic investment and only requires some up-front "how to" education and coaching that is readily available from multiple sources. For those who prefer using some external expert assistance, the amount of assistance really needed is minimal. While it's easy to view VSM as only a procedural methodology because it has well-established process steps, VSM is much more of an art than it is procedural—while there are defined steps, the art comes into play when interviewing people, looking for opportunities, designing the future state.

But, the gathering of accurate data is critical and the preference is to get them firsthand by observation and the use of the old-fashioned stopwatch. While the use of historical data from corporate information systems at first seems logical, experience in the Lean community is that it isn't—all

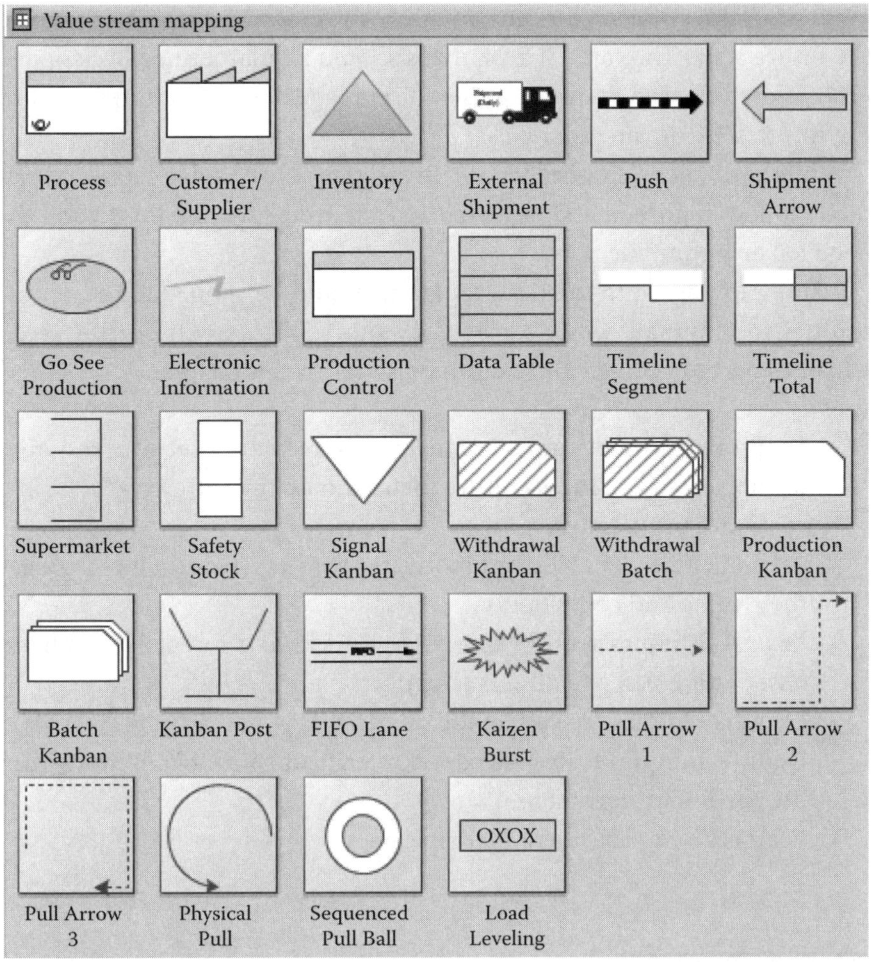

FIGURE 7.8
Additional value stream mapping symbols. (From edrawsoft.com. With permission.)

the definitive books on the subject specifically say do not take information out of the legacy information system; you *must* get up, go out, and see!

Those educated and trained on VSM should include everyone involved in management and supervision. Everyone within this extended team may not develop a value stream map but needs to have confidence that it is the road map and prioritization tool for authorizing kaizen events (covered below) and other improvement activities (i.e., projects and just-do-its).

Senior management should engage in the actual mapping events because of the strategic nature of VSM, and a VSM should be developed for every major product family and all the administrative, manufacturing processes, and information flows within them for both the current and the future state. The value stream maps should be maintained by assigned process owners, and reviewed by senior management for progress toward the future state on a regular basis.

While a value stream map doesn't directly involve possible capital investment, it does highlight and prioritize kaizen events that might involve the need for new equipment, etc.

Before moving on to the Lean tool that should be applied to the opportunities to eliminate waste surfaced by the VSM, we will quickly recap what has been discussed and examined so far in this chapter:

- Lean or continuous improvement is more than a set of techniques or tools. It's a business culture focused on eliminating waste from a customer perspective.
- A Lean culture enables and motivates workers because it recognizes their value and contribution.
- Lean techniques require thorough clinical and impartial firsthand observation before action is taken.
- Lean techniques are systematic.
- Lean techniques foster and develop critical thinking ability within the corporate workforce.
- Lean is an improvement machine.

KAIZEN EVENT OR KAIZEN BLITZ

Technique Overview

Kaizen is a Japanese word that means "continual improvement." Translations are interpretive, and the literal transliteration from Japanese to English reads *kai* = "change" and *zen* = "good," hence *kaizen*. What is the purpose of a kaizen event?

> A two to five day focused improvement activity during which a sequestered, cross-functional team designs and *fully implements* improvements to a defined process or work area, generating rapid results and learned behavior.
>
> **—Karen Martin and Mike Osterling, *The Kaizen Event Planner*, 2007**

The targeted improvements are focused on the elimination of non-value-adding activities, with aggressive and measureable objectives, and working within the boundaries of existing resources. This is in contrast to a project that typically takes more time to complete, and often more capital equipment and human resource investments. Another contrast is the simple kaizen (to be discussed next), which is a "just do it" improvement that takes a few hours or less to implement and doesn't necessarily require cross-functional teaming.

A kaizen event and a quality functional deployment (QFD)/outcome-driven innovation (ODI) study both share a common requirement; they both require a cross-functional team to satisfy the need for "eyes" with different perspectives to see the often hidden opportunities for improvement, and ideas of how to implement them. Part of a cross-functional team is the event facilitator, who can be an employee or an external resource. Many companies will use an external facilitator from time to time because they have highly developed facilitation skills and also function as a teacher/mentor.

What identifies the need for a kaizen event? There are two things. One is the most wasteful process identified in the current-state value stream map (VSM) that increases throughput within a process. The other is identified safety issues, or task ergonomic improvement opportunities, that if not acted on can cause harm to worker-associates.

A kaizen event that meets its targeted improvements is an event that is carefully planned. Planning includes just-in-time education of the kaizen team members if they are new to the application of the kaizen process or the Lean techniques. The important considerations in preparing a team include:

- A cross-functional team with members from different disciplines in the organization. Personnel who have never seen a process usually ask the best questions on why something is done. This helps to motivate new thinking among team members.
- Using common and shared measurements; they function as a shared language.
- Preparing them on how to properly take measurements to ensure accuracy and establish a workable and credible baseline.
- Helping them understand that there is a need for viewing what is going on accurately and objectively, and our natural tendency not to see the obvious. Yes, individuals can be trained to do this. A favorite video of experienced kaizen facilitators is one in which a number of individuals are passing a ball to each other. A person dressed in a

gorilla suit walks through the group during the ball-passing exercise. During the first viewing of the video many observers don't even see the gorilla, and those that do can't recall how often it passed through the crowd. It's a great exercise to get people focused on the need to view how work is done carefully.

- Helping them understand that the whole is greater than the sum of its parts. That a collaborative outcome is almost always better than that of an individual, despite that person's considerable talent.
- Improving each team member's listening ability and the responsibility each has to not make assumptions and to confirm he or she understands what was intended in a statement by another team member. A good technique to achieve this is for someone to paraphrase what he or she thought the person said, "If I understood you correctly, you suggested XYZ." If that paraphrase is not completely accurate as the speaker intended, he or she now has the opportunity to refine the communication. Seems silly, but consider that English is a rich language that can often unintentionally be abused. Research has shown that fifty percent of what is said in a conversation is not understood in part because both the speaker and listener made assumptions and didn't have a convention to clarify.
- Establishing team etiquette. This is particularly important with groups of people who haven't worked in a Lean culture before. They will soon learn that respect for the individual is a cornerstone of Lean.

A brief overview of the preparation necessary to conduct successful kaizen events isn't really adequate, and the reader is urged to read the *Kaizen Event Planner* guide mentioned earlier.

The kaizen event is the workhorse of Lean/CI and is figuratively a Swiss Army knife for elimination of waste. It is industry neutral, working in any industry, from food processing to service providers, including government! Organizations that are mature in their pursuit of Lean/CI always seem to have a kaizen event going on somewhere in the organization, week in, week out.

Management Considerations

There is a planning activity that precedes conducting the actual kaizen event. The improvement to be achieved along with its targeted results must be established. How much improvement is planned should be ambitious but achievable. An example might be to increase throughput within a work area by fifty percent. Another pre-event consideration is the budget for implementing the changes required to achieve the targeted improvement, but the emphasis of events must be on creativity before capital—and in fact, most events will have a very small budget. The goal is to have a fast return on the investment from the changes and a sustainable change. One of the benefits of the Lean management system is that it is self-funding from the improvements realized from using CI techniques such as the kaizen event. The management responsibility is to review the kaizen event plan for its integrity, and if it meets the following criteria:

- Improving the performance of a major constraint identified in a value stream map that is consistent with the planned future state.
- Has appropriate improvement targets proposed.
- Has an acceptable funding request.
- Accountability has been established along with who will be facilitating the event.
- A plan to do just-in-time education and training of those involved in the event has been developed, and the identification of an external facilitator, if one is appropriate, has been made.
- Arrangements have been made to minimize the disruption from the event on normal production and business flow.

If all the criteria have been addressed in the plan, management should authorize the kaizen event.

There are other management considerations. The development of a time-phased master kaizen event plan makes sense. Corporations run production from a master production schedule (MPS), so why not have a master kaizen event plan derived from the future-state value stream map?

A further consideration is sustainability of the improvements made through a kaizen event. These should be monitored for an extended period of time. How long that should be is a judgment call, but the monitoring and reviewing should be formalized.

A final point, a kaizen event must contribute in a significant way toward achieving the future-state VSM. A whole earth view of the opportunities for improvement is necessary. An opportunity that looks compelling and acted on without considering its impact on downstream activities could create new unanticipated and unwanted bottlenecks.

KAIZEN

Technique Overview

Because a kaizen event (blitz) and a kaizen share the root word *kaizen*, the new student of the Lean system can be confused. Perhaps two very distinctive titles would be better, but as they say, it is what it is.

> A kaizen is an improvement that generally is recommended by one individual or a small group within a work area. Often the improvements are small, but with others cumulatively related improvements create significant value. They are spontaneous in contrast to a structured kaizen event. Those experienced with kaizen are skilled at offline testing and experimentation without impacting current methods of performing work. Some of these skills are acquired through participation in formal team-oriented kaizen events.
>
> **—Mike Osterling, Osterling Consulting**

Management Considerations

The ideal situation is for a group of people who are intimate with the work being performed to make changes that eliminate waste. Proficiency at effective kaizen requires experience from doing, and that is best acquired through managed kaizen events. A team of workers capable of performing kaizen often become self-managing, a situation that is very rewarding for them and their now nominal manager-coach.

Because of food safety considerations, it might be prudent to have all suggested changes quickly reviewed by internal food safety practitioners before implementation.

Generally there are no major capital expenditures required because of a kaizen. Also, there is minimal education and training expense because it has been absorbed in preparing for the initial kaizen events the individuals participated in. After that the education is a by-product of doing work.

5S/6S

Technique Overview

6S is focused on eliminating some of the sources of waste through simplification, creation of order, visualization, and making problems easy to see.

The influence of 6S is already partially present in food and nutraceuticals manufacturing in the form of sanitation standard operating procedures (SSOPs).

6S goes beyond that in that it is focused on ridding the entire enterprise of clutter, unnecessary items, and improving safety in not only manufacturing, but also all the work areas, including supporting functions such as administration, recipe development, etc.

Why is 6S needed? Nearly everyone has pressure to get work done on time, within budget, and operates against tight schedules or standards. Without the guidance and discipline that 6S brings to workers, clutter builds up in a workspace. With clutter comes confusion, hunting for tools or parts or instructions, etc., which all contribute to creating waste and perversely makes it harder to perform to expectations, and a vicious cycle feeds itself. Waste almost seems to have reproductive capabilities—it breeds! Humans are adaptive; we can adapt to most circumstances, and in the case of clutter and confusion, it soon becomes normal, even if uncomfortable to deal with. The waste that clutter creates is amplified in work areas that have more than one type of process performed and are shared by coworker teams. 6S is really focused on the process and not the workplace, but the workplace is where the positive results are most visible. The key elements of 6S are:

- Sort: Distinguish between what is needed and not needed and remove the latter.
- Stabilize: Enforce a place for everything and everything in its place.
- Shine: Clean up the workplace and keep it clean.
- Standardize: Monitor and maintain adherence to the first three Ss. In addition, standardization includes visual controls to help manage the business.
- Sustain: Keep the workplace 6S compliant; maintain the gain.
- Safety: Eliminate safety hazards.

Note: Safety hazards and poor process step ergonomics are also frequently targeted by kaizen events, which were discussed earlier in this chapter.

Because the focus of 6S is on the process, not the workplace, it is important that if a workspace supports a variety of processes, one process is not optimized at the expense of others; all must be supported. If that is not possible, one alternate strategy is to create a dedicated cellular work area for the process that is creating the maximum value. Teams and cells will be discussed later in this chapter.

Implementing 6S requires the same considerations for managing change in general, except it is collaborative. As with most Lean techniques, direct observation is the key. It is important to not use secondhand information or abstracted information. A flexible implementation scenario follows:

1. Communicate or introduce the value of 6S to senior management and secure genuine buy-in. Make sure that they understand that they need to be seen as physically supportive, and this isn't just another improvement crusade.
2. Communicate or introduce the value of 6S to work area leaders in the organization. Sometimes this takes a few cycles of communication for individuals to get it. People learn in layers that progressively build up knowledge and understanding. This is really the adult learning principle 101.
3. Focus first on those who get it. These are generally individuals that are dealing with work area problems and are inclined to be early adopters of new and better ways of doing business. Their energy and enthusiasm will create early successes that will be a perpetual promotion machine for 6S to other work area leaders.
4. Confirm that each 6S event makes sense.
5. Establish goals and ground rules for the event and get buy-in from the work area leader and process owner.
6. Communicate or introduce the value of 6S to work area workers who will collaborate and contribute to the transformation. If possible, let their leader do the introduction. Present the simple step-by-step process that will be used and the prerequisites and equipment. Paint a positive picture of the future state and how it benefits them.
7. Document the current state on the 6S evaluation form (see Figure 7.9).
8. Organize the materials and equipment that will support the event.
9. Do it!

Production Line: ____________

Cell Number: ____________

Date: ____________
Auditor ____________

Audit Point	0	1	2	World Class 3	Score	Observations / Items	Target?
Sort							
1. Are all items in the areas used on a regular basis?	>10 items found in the area that are not used	5-9 items found in the area that are not used	<5 items found in the area that are not used	Only used items are found in the area			
Set in Order							
2. Do all items in the area have a dedicated location and is it stored in that location (unless being used). Check for hand tools, materials, WIP, scrap, documents, supplies, etc. Look at work stations and in common areas.	No evidence of implementation of any points	Implemented in < 50% of the area.	Implemented in > 50% of the area	Implemented in 100% of area and disciplined adherence			
3. Are all items in the area identified with description and home location?	No evidence of implementation of any points	<50% have either home location or description	>50% have both home location and description	100% have home location and description			
4. All tools, supplies and parts are visible?	Drawers, closed cabinets and toolboxes are in use.			Any closed storage has clear covers and dedicated locations within.			
5. Are department identified and work boundaries clearly marked?	No lines or signs	Only one of two identified		Lines and signs exist			
6. Are emergency equipment (alarms, extinguishers, electrical, panels, etc.) and evacuation routes identified?	Not 100% compliant			100% compliant			
7. Are ESD areas and other PPE requirements clearly identified and respected?	Poor visuals or compliance			Visible and compliant			
Shine							
8. Are cleaning responsibilities defined (who, when, what) and posted?	No standards in place	Standard defined-visual or verbal - not followed	Verbal standard in place, being followed.	Visual standard in place and being followed			
9. Is the area clean? Check racks, equipment, floors. Is there dust, oil, debris?	No evidence of regular cleaning	Little more than sweeping appears to be done	Equipment, floor and benches are clean—stains on the floor	All surfaces are freshly cleaned.			
Standardize							
10. Is competency matrix and OJT status posted in the area? Check 3 operators.	Doesn't exist	Exists, but not visual and not updated	Updated but not visual or not updated	Matrix updated and visual			

FIGURE 7.9
Monthly 6S production audit form. (Courtesy Osterling Consulting. With permission.)

10. Conduct scheduled audits on a frequent basis to maintain this new desired state. A sample 6S audit checklist is shown in Figure 7.9.

Since the corporate goal is continuous improvement, many, if not all, work areas will have at least one kaizen event to improve processes and many more over time. The outcome of the kaizen event may change the current 6S configuration for a process in a work area.

Management Considerations

There is an up-front educational expense needed to implement 6S. Education is needed to convey the hidden cost of a disorganized work area because people quickly become acclimated to confusion and think it's normal. Teaching a group of people about the benefits of 6S has two threads: the cause of confusion that creates waste, and visual examples of disorganized work areas in contrast to organized work areas. Most of this expense is because the education is conducted outside of normal work hours or briefly consumes production time. Generally this is a modest expense.

Another consideration is leadership by example. Management must take the initiative on 6S and prove it by its actions, in some cases actions as simple as:

- Ensuring that the corporate reception area is organized, spotlessly clean, and gives visitors insights into the corporation's competitive strengths
- Keeping conference and meeting rooms organized, clean, and equipped with equipment to support effective group communication and collaboration
- Maintaining a pleasant employee luncheon and break area whose design shows respect for the employee
- Keeping the property around the corporate building clean and organized
- Conducting the critically important weekly activity—6S audits of work areas

A brief recap of what has been discussed so far follows:

- PDCA (plan-do-check-act) and A3 are techniques that refine problem-solving and thinking abilities focused on eliminating waste.
- Voice of the customer (VOC), such as quality functional deployment (QFD), and its evolutionary offspring, outcome-driven innovation (ODI), are methodologies to acquire a deep understanding of what a customer truly values and is willing to pay for.
- Value stream mapping is the technique to analyze and visually depict the constraints in a process. This big picture, in combination with the insights acquired through VOC studies, provides focus beyond the normal objective to improve productivity and business process velocity. A value stream map helps identify the areas that should be attacked with a kaizen event (blitz).
- The kaizen event is figuratively the weapon of mass improvement to work toward the desired future-state processes.
- 6S is focused on creating an orderly, uncluttered physical environment that in turn helps create healthy thinking. Using a health analogy, a healthy body helps maintain a healthy mind. Many organizations start their Lean journey with 6S, and once again an analogy is appropriate: 6S is the warm-up and stretching required before an athletic activity.

We are at a difficult point in the discussion, and the age-old riddle of what came first presents itself: What came first, the chicken or the egg? Developing a team collaborative spirit and establishing accountability, and organizing the workplace into a cell configuration that visually communicates to its members and other work areas aren't dependent on discovering opportunities to eliminate waste through a value stream map or the subsequent kaizen events. Workforce development and well-crafted industrial-manufacturing engineering are basic best practices to improve workforce productivity and gain some competitive advantage. Unfortunately these core truths have been forgotten by many. The good news: The Lean system once again elevates them and makes them a priority.

The key thought is that putting emphasis on laying a foundation in advance of team competence, and workplace organization will yield better results from kaizen events than organizations that haven't done so. An athletic analogy: A person has trained for and run numerous 10K races and wants to run in an important marathon. Another person also wants to race in the same marathon but has never trained for or participated in a running event. Both are equal in terms of health, athletic ability, and tenacity. Which one would a betting person put his or her money on?

WORKPLACE ORGANIZATION

Technique Overview

For food and beverage and nutraceutical manufacturers there already exists guidance on workplace organization. Both operate under the U.S. current good manufacturing practices (cGMP) and Occupational Safety and Health Administration (OSHA) regulations. In the European Union (EU) and other regions in the world there are similar requirements to those in the United States. But these by themselves don't provide enough guidance to achieve a Lean, efficient, and rational workplace.

While it may seem logical that a vigorous first step would be 6S, it would be a mistake, and many companies make this mistake. Again, what seems logical isn't. Because if they start by applying a tool (6S) before they have defined or understood what the problem is that should be addressed, they will needlessly expend energy (a waste), and put in question the benefits of the Lean management system. VSM is a much better starting point, as it is focused on identifying issues as well as appropriate tools/techniques to use to address those issues.

Another is the adoption of standard work, which will be discussed below. One possible supplemental approach is moving supervision and management when possible into the work areas and not letting them hibernate in an office that isolates them from the action in the workplace. They really don't want to be distant from the action; traditional conventions and a top-down culture have mindlessly required it. Leadership needs to be in the center of those being not managed, but enabled.

Aligning the organization along product families (including manufacturing cells) versus generalized work areas often makes sense. Anytime work has to travel any distance it creates waste in time and handling and pulling together equipment to perform the work.

In situations where there is a high level of automation, thinking through how to eliminate waste by organizing machinery and people has already taken place, and in effect, it has a cell-like capability. Often automated processes with more than one piece of equipment are a mismatch of processing speeds and capabilities. The advantages of organizing into cells when possible will be discussed next. Another goal in organizing the workplace is making it visual; no one should have to go to a computer screen to understand what is going on in any work area. They should be able to glean that by just walking through it. Easier said than done, but achievable with a continuous improvement perspective, and with experience it gets easier.

TEAMS

Technique Overview

The importance of teams has been mentioned in earlier sections of this chapter. Because of the importance of people effectively collaborating together in teams to achieve the goal of continuous improvement (CI), it makes sense to emphasize the need to develop a culture of teaming.

Achieving a collaborative culture for some may require a change in beliefs. In the United States, for example, we've been immersed in the belief that the independent individual, infamously called the Marlboro Man in the past, is critical for the well-being of society and all of its component pieces. Individuals with a solid internal compass on matters relating to ethics are important and critical to civic fairness, but extreme individualism in the context of problem solving doesn't produce results that will move forward the goals of CI. Harnessing the collective power of the team working collaboratively will. In an organization that has a strong Lean/CI culture, individuals have learned how to balance the positive aspects of individualism with the ability to collaborate with others effectively. A facilitator of this learning process is the development of respect for others that evolves into trust.

Another important element is the deliberate but patient education process that introduces individuals to the benefits of group collaboration, lets them internalize what this means for them and the organization they work with. An excellent way of internalizing is through a game rooted in reality that is fun and thought provoking. One game that comes to mind is a desert survival one that is fact based and dramatically shows the advantages of team collaboration over individualism. The game:

- The participants are the survivors of an aircraft that has crashed in a harsh remote desert landscape.
- Teams of four to five participants are formed. In a corporate setting there are normally multiple teams.
- After the crash they believe they are X miles/kilometers from water and a place that will help them survive. The distance they must travel is challenging given the harsh environment.
- The aircraft has some supplies that might be useful in helping them reach safety.
- The supplies are described by the game facilitator as well as some general strategy choices.
- Each individual on a team is asked to privately make choices on the supplies they will use and a strategy to survive. There are practical limitations; utilizing everything is not practical. The individual results are not shared with team members.
- Each team is then asked to discuss and collaborate on the strategy and equipment choices. There is give and take, the stakes are high, and their lives are at stake.
- All the results are gathered up and compared to the strategy and equipment list based on actual choices that resulted in survival or death. For everything that didn't contribute to survival there is an explanation as to why.

The result? Almost all the choices made by individuals resulted in failure, and most group-developed choices resulted in success.

There are many different exercises and thought-stimulating games that will help individuals understand the reason for and value of team collaboration, both to them and their team and organization.

ADVANTAGES OF THE MANUFACTURING CELL

Technique Overview

Manufacturing cells are designed to connect a series of (often dissimilar) process steps to support the production of a specific value stream product family. One goal of a cell is to minimize the handling and time from process step to process step within the overall process. A well-designed cell accommodates the "pacemaker" in the overall process. The pacemaker is the series of process steps, closest to the customer, that produce at the rate of customer demand.

An advantage of a cell is its dedicated equipment and people. These trained workers are the real instigators of improvement because they are intimate with each process step. The flow of work is an important consideration in the design of a cell. It should support the shortest processing time and product movement and consume the smallest amount of space possible. Lean is frugal! One common configuration is the U, with work entering the U and being processed or flowing within it on the inside and exiting the other side of the U (Figure 7.10).

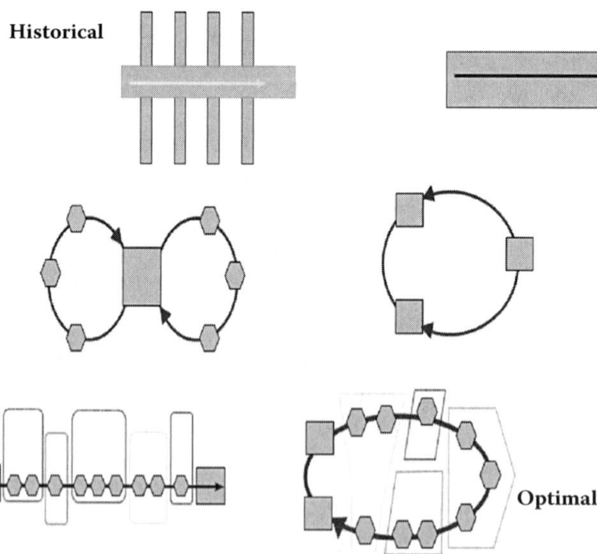

FIGURE 7.10
Examples of different equipment-people-process configurations to support the completion of work. (From Ron Calame, California Manufacturing Technology Consulting® (CMTC) Torrance, CA. With permission.)

The U configuration has the advantage of a worker performing an operational step and turning around, walking a few steps, and performing a subsequent process step. Workers in a cell are often cross-trained to perform more than one process step and also trained to work in other cells. This provides flexibility in moving the worker team to the point of greatest need. Lean does emphasize smaller batches of work when possible, and a cell team will frequently reconfigure itself to address the immediate circumstance.

A manufacturing cell doesn't necessarily need to be fixed in a permanent location and configuration. One approach is to have all equipment on rollers or wheels with electrical and pneumatic connections in the ceiling. When a production batch needs to be processed, the needed equipment is rolled into a cell configuration and connected to the grid overhead.

When the production batch is completed, the equipment is disconnected from the grid and the equipment is available for other needs.

One might think that these flexible islands of production capability might be unmanageable if organized in any large number. This is not correct. I know of a company with over 120 cells that it efficiently and easily supports:

- Processing a large number of items within numerous product families
- Two thousand employees
- Three shifts per day
- Seven days a week

Considering the high-mix, high-volume output, the manufacturing area footprint is very small (frugal), productivity is high (and increasing), compliance to the schedule is high, time to process from start to finish is short, and the velocity of the product through the enterprise keeps on increasing. One of the tenets of the Lean enterprise is one-piece flow, and a cellular approach supports that goal.

If a company has invested in high levels of process automation, it has hopefully achieved the benefits of process optimization, but there is still the likelihood that there are other processes that are candidates for the manufacturing cell approach.

The majority of food processors are contract manufacturers that might find it difficult to invest in high levels of automation to reduce costs because this requires significant capital investment. Investigating cellular manufacturing might be a very good, cost-effective alternative. Even their customers, the brand owners, should be coaching their small- to medium-sized suppliers to do so.

If someone wants to learn more about designing and using the manufacturing cell approach, there is good news. There are many books written on the subject, and the Association for Manufacturing Excellence (AME) has put together educational programs on this approach. Also, AME has encouraged companies who have had success with any Lean concept, methodology, or technique to actively share the lessons learned with other companies. AME calls this the company-to-company connection.

The cellular approach isn't restricted to manufacturing processes. It can be applied in select situations to many engineering and administrative processes. The toolkit of Lean techniques is large and flexible and, for those who understand their underlying logic, useful in improving any process.

VISUAL MANAGEMENT

Technique Overview

The visual workplace is difficult to describe in words because it's visual! There are some basic objectives in the journey to achieving a visual workplace:

- Anyone walking through a work area should know what's going on in that area, without asking those performing work or going to a computer terminal to determine the performance of the area, what is being worked on, and its current problems requiring assistance and intervention.
- Any area that supplies another should know through visual signals what its "customer" needs from it.
- In the event of a problem impacting a work area's ability to perform to expectations, it is visually communicated to those who can assist in resolving the problem. One common approach is the andon light. An andon light is similar to a traffic light: Green is go (everything is OK), yellow is a caution (a transition to red is occurring), and red means stop. Red on an andon light is a signal for help.
- Everyone in a work area should know how the team is performing toward goals, and this information is posted in a central location within the area. The status is presented not only in numbers highlighting variances, cumulative achievement, units required to meet the goal, etc., but also graphically. There are typically multiple time

periods: today, this week, this month, and this quarter. All material shortages and their expected replenishment dates should be posted, and if an andon is red, what is the problem that needs to be resolved.

In keeping with anything involved in continuous improvement (Lean), there is never an endpoint in the evolution of the visual workplace, but a desire to share important information in an intuitive way to coworkers and others in the enterprise. Once an initial foundation is laid down, the improvements in visual communications will come from the workforce if their contribution is encouraged and acknowledged. There are two ways to get rich. One is to hit the jackpot and get one big reward. The other is to get frequent dividends on wise investments. One word aptly describes the visual workplace; it's an *annuity* that keeps on paying and it's essentially cost-free.

Management Considerations—Workplace Organization, Teams, Manufacturing Cells, and the Visual Workplace

There are at least six significant considerations:

1. Attitude adjustment in the Lean system executives leads and enables others to be successful. One good way of visualizing this is to take the traditional organization chart and turn it upside down.
2. The need to develop a culture of mentoring. This often starts with the development of a small team of individuals who are excited about the Lean system, want to learn more, have some natural people skills, and enjoy helping others do a better job and increase skills and contribution to the organization. In the Lean system those who mentor or teach are held in the highest esteem. In the early stages of adopting the Lean management system some organizations bring in individuals who have demonstrated success in the past with other companies to "seed" the organization with this mentoring mindset.
3. Understand the need for patience and then develop it. Changing a corporate culture takes time, particularly for organizations that historically have used a command and control style of managing. Those textbook success stories on the Lean system tend to be on companies that have been on the Lean journey for at least three to five years. This doesn't mean that it takes years to see a return on investment!
4. Having a solid industrial and manufacturing engineering team that buy in to the Lean system is important. Unfortunately for some

reason the appreciation for these two disciplines has declined over the last few decades, but their contribution is significant. They design the platform upon which the rest of the enabled organization can continuously improve.

5. Conduct a pilot on the approach to adopting Lean cells. This provides a learning experience and minimizes risk.
6. There will be investments in workforce development and, on occasion, the need for capital investments, but all will be easy to evaluate. The mantra within the Lean community is to simplify, simplify, and simplify again until the need for capital investment is gone or its need highly obvious and justified.

STANDARD WORK

Technique Overview

One of the purposes of standard work is to design repeatable processes by removing variation, creating consistency and the ability to produce at a standard rate. Standard work is formally defined and documented. Generally speaking, standard work encapsulates the sequence of tasks (operations) done by a single worker or a team coordinated in adding value in a task. When implemented it provides a stable and reasonable working environment. Another is to create a foundation for future improvement.

> Where there is no standard there can be no improvement.
>
> **—Taiichi Ohno (a revered Lean system innovator-philosopher)**

The building blocks of standard work are standards. Standards establish the rules for how something is done. Examples of standards are workplace safety (U.S. OSHA), quality standards, food safety guidelines (cGMP), internal standards (SSOPs), expected rate at which work is to be performed, etc.

What can be incorporated into the standard work for a task is flexible. The considerations are the complexity of the task, the skills of those performing it, and the available technology to support it; common sense prevails. The standard work document might include:

- Descriptive text, supplemented with the below
- Target values, both performance and quality
- The equipment required
- Safety considerations
- Regulatory boundaries (food safety)
- Graphical depictions
- Videos
- Auditory support
- Multiple spoken and written languages

While the noun *document* evokes thoughts of a written document, it also now includes other components:

> - An object containing information . . . such as a movie, photographic, or audio recording
> - A computer file . . . using a database, spreadsheet, figure, or text file
>
> —***Encarta Dictionary*, English (North America)**

While this might sound futuristic, it's not. The smart mobile phones and computing tablets of today, e.g., the iPad, make economical delivery of a standard work document at the point of use easier and less expensive than ever. Creating the standard work document utilizing all these different communication mediums is readily available today even to the private consumer.

Standard work is a key building block that interacts with 6S, workplace and cell design, the visual workplace, team dynamics, and kaizen events. All these Lean tools collaboratively push the continuous improvement mandate forward.

Products manufactured under Hazard Analysis Critical Control Point (HACCP) controls are in essence produced using standard work, particularly the critical control points (CCPs). Utilizing all these interrelated Lean tools will create a superior food safety manufacturing environment *and* improved productivity to compete in the new global economy.

Management Considerations

Standard work is logical workflow instruction, with accurate and complete data, that integrates all the information needed to reliably make a quality product every time. Unfortunately in many companies, instructions and data to support all aspects of manufacturing a product are fragmented, which can contribute to food safety risks, and also to suboptimal organizational productivity. This fragmentation of information generally has occurred over an extended period of time, and most individuals in the organization have acclimated to it, and for them it's normal. Some data are on the legacy computer system, and also on electronic spreadsheets, some of them reside on paper-based documents, and much data are in someone's head.

Fixing this fragmentation requires centralizing and integrating all the data electronically, and presenting them as a process flow. When data are organized in this manner, the disconnects in information flow, and what's needed to correct them, become transparent and obvious.

Centralizing data to create world-class standard work is a project. Why not use it as an opportunity to pull all relevant data together and create a repository of integrated master data? A project of this type requires investment in education, manpower, and potentially in some data processing equipment or software.

The know-how to bring data in different formats together, prune the unnecessary, and organize it into a cohesive whole exists today within the information technology profession; it just needs to be applied and is often referred to as systems thinking. If a company doesn't have staff with these skills, it will have to recruit them or invest in existing staff to master the techniques required. Also, procedures will need to be developed to maintain this integrated master data.

The possibility of investment in data processing equipment or systems was mentioned earlier. One such investment would be the implementation of a process batch ERP system to centralize data. Another is to make data easy to transmit, and finally make access to information available in a usable and natural way. And, do all of this on a cost-effective basis. In the past this was daunting to those managing a small- to medium-sized company. Today that is not the case; we are currently going through a time period that can best be described as disruptive when compared to the use of traditional data processing equipment and information technology. Information technology has evolved to the point where it is powerful, cheap, and useful:

- Powerful: Over thirty years of evolution of the microprocessor has created handheld consumer devices with the power of mini-computers (think 1980s).

- Cheap: A cell phone or tablet device is a powerful computer produced by the millions at a consumer (versus industrial) price point.
- Useful: Convergence is the reason for usefulness.

Convergence of data, graphical user interface (second generation), voice, vision, bar coding, and video into one integrated device whose use has been mastered by the incoming generation of workers.

A side note: There is another management consideration for the current generation of senior managers when new technology is introduced into an organization, the need for cross-generation mentoring. Many executives today over forty years old are information technology challenged, but experienced and knowledgeable on best practices, and how business flows come together to produce a positive result. Those in their twenties would benefit from this experience and expertise. They are the next generation of executives. For this generation information technology in its converged state is second nature. The challenge is to set up two-way coaching relationships.

QUICK CHANGEOVER—SINGLE-MINUTE EXCHANGE OF DIE (SMED)

Technique Overview

The origin of quick changeover, or as it is often referred to, SMED, was the automotive industry, specifically Toyota Motor Company. Setting up for a production batch often requires equipment and tooling changeover. The cost of changing over can be very expensive. The solution was to increase the batch size, but this increased inventory investment. Inventory for most industries within manufacturing is the largest cost component of operating, and any excess inventory is expensive, despite antiquated accounting rules; excessive inventory is pure waste, not an asset.

The goal of SMED is to reduce the size of production batches, which in turn reduces inventory investment. The SM in SMED means single minute, and the goal was to reduce equipment setup to be a major consideration in calculating batch size. If it only took one minute, then I could just make what I needed with no economic consequences. Great goal, often difficult to achieve, but what did develop that was easy to achieve was thoughtful preparation to minimize the changeover process time:

- Separate external elements from internal elements.
- Every tool needed is reserved and put in the correct sequence of usage.
- If materials are needed they are in the changeover kit.
- Tasks in the changeover that aren't at the highest skill level are assigned to an apprentice.
- A specific changeover is rehearsed and refined based on experience. Practice makes perfect. A good example is the automotive racing series NASCAR. NASCAR races are often won and lost on the time it takes for tire changes, fluid refills, and minor repairs in the pit area during stops during the race. Pit teams practice, practice, practice all these procedures, and then practice some more!

As with many best practices, they are often developed in industries outside of one's own home industry. A famous jazz musician once said during a concert, "Why is it that if you steal the thoughts of one person it's called plagiarism, but if you steal the thoughts of many it's called research?"

The author of this book encourages the reader in the food and nutraccutical industry to heavily engage in research.

Management Considerations

It's a given that adherence to cGMP is priority one, but there will be opportunities within those boundaries to streamline and shorten changeover times.

Learning how is easy. There are formal courses on the key elements and application of SMED, and it's a small investment that over time and on a compounded basis increases production productivity.

Process Batch ERP

Applying and fully benefiting from the application of CI techniques requires an operational baseline that is stable. The baseline doesn't have to reflect world-class inventory management or industry-leading productivity, but it must be stable, that is predictable. Process batch ERP provides that stable foundation. This doesn't mean that an organization has to have process batch ERP implemented to start on its implementation of Lean. Many small food processors don't, and many larger firms that do aren't utilizing ERP fully. But in the big picture scheme of things, process batch ERP is an anchor needed for stability.

SUMMARY

Lean is a management system and organizational culture that is focused on delivering maximum value to the customer, elimination of waste in all forms, and enabling the organization's workforce. Its cornerstone is respect.

There are numerous techniques to support the Lean management system, which is commonly and collectively referred to as continuous improvement (CI). This chapter covered some of the CI techniques, with a brief overview of each, followed by its management considerations. The CI techniques cluster into five groups:

1. Analyzing, critical thinking, and problem solving (PDCA and A3)
2. Getting in alignment with customer needs (VOC, QFD, and ODI)
3. Identifying opportunities for improvement (eliminate waste) within process flows (spaghetti diagramming and value stream mapping)
4. Organizing and the removing of major waste (kaizen event or blitz)
5. Creating and maintaining a Lean infrastructure (6S, workplace organization/cell/visual, standard work, and SMED)

Also covered were the relationship and interaction between Lean and process batch ERP.

QUIZ

7.1. What is Lean?

7.2. What are the four primary objectives of Lean?

7.3. What is the relationship between Lean and continuous improvement (CI)?

7.4. Implementation of Lean requires extensive investment in capital equipment.
☐ True ☐ False

7.5. Which technique supports the PDCA cycle: 6S, A3, or Kanban?

7.6. The A3 method supports visualization of a problem and the proposed resolution.
☐ True ☐ False

7.7. What technique systematically surfaces customer needs?

7.8. What quality permeates an organization with a Lean culture?

7.9. What is the primary purpose of 6S?

7.10. What analytical tool objectively and systematically identifies the status and performance of each node within an end-to-end process?

7.11. What is the goal of a kaizen event/blitz?

7.12. What are the two methods used to identify potential kaizen event candidates?

7.13. Kaizen events are generally spontaneous.
☐ True ☐ False

7.14. What are two differences between a kaizen event and a kaizen?

7.15. Name a few concepts or techniques used to guide the organization of a work area and their evolution.

7.16. What is the relationship between food safety procedures and standard work?

7.17. Material requirements planning (MRP) is the only proven material planning and control technique available.
☐ True ☐ False

The answers to these questions can be found in the Appendix along with definitions, website addresses, and recommended reading.

8

Sustainability

> Humanity has moved from using, in net terms, about half the planet's biocapacity in 1961 to over 1.2 times the biocapacity of the Earth in 2002. The global ecological deficit of 0.2 Earths is equal to the globe's ecological overshoot.
>
> **—Arjen Y. Hoekstra, *The Water Footprint of Food***

EXECUTIVE OVERVIEW

This is an important chapter for those participating in the food ecosystem—suppliers of raw materials, food processors, nutraceutical manufacturers, those who store or transport product, and the food retailer. The food industry is energy and water intensive, both of which are becoming expensive, and in some regions of the world in short supply. Consumers and governments are increasingly concerned about the health of our ecosystem and its sustainability.

All food products require water. Many require heating and refrigeration. Both of these commodities consume energy and, using current sources of energy supply, create unwanted carbon emissions. In addition, for highly centralized food production, logistics is a major component of doing business. Logistics is energy intensive.

Given these constraints, a major rethink is warranted, "change is a coming," and a major U.S. FDA report confirms the need for frugality in the use of resources:

> In the decade ahead, the world economy will be shaped by several distinct forces: the rise of emerging markets, *the scarcity of natural resources*, and the increased flow of capital, information and goods across borders. [emphasis mine]
>
> —***Pathway to Global Product Safety and Quality*, FDA**

In previous chapters of this book, we've discussed the advantages of process batch ERP, the Lean management system, and how effective supply-chain management contributes to good food safety practices. All three also contribute to the elimination of waste. An excessive carbon and water footprint is the most onerous waste of all.

The objective of this chapter is to have the reader add an additional perspective to his or her thinking, and to consider every aspect of his or her corporate environment and implications for its carbon (CO_2) and water use. Two points that should be clarified before we continue:

1. When CO_2 emissions are discussed in this book, this is shorthand for all six greenhouse gas (GHG) emissions contained in the landmark Kyoto GHG Protocol accord.
2. The term *sustainability* has a rather fluid meaning, and to some people connotes the higher-level concept of the triple bottom line, which in a balanced scorecard fashion attempts to balance people, the planet, and profit (the 3Ps).

The good news is that the methodologies and techniques to identify and reduce carbon generation, and minimize water usage, are maturing to the point of being usable. While sustainability is, in relative terms, still in its early stages as a "must-do" business imperative, and there is significant competitive advantage to committing to it now, it is very likely that current and future customers, and their customers, will demand sustainability actions from their suppliers in the near future.

INTRODUCTION

Concern for our ecological environment is now embedded within most societies. This concern is based on the real or perceived degradation of the environment, and the growing scarcity and cost of important raw materials. There are many contributors to this concern:

- The explosive growth of our global population
- Dealing with the unintended consequences of earlier industrialization
- The rapid industrialization and urbanization of many countries within the third world
- Better application of science to understand the physical and ecological consequences of our actions
- The emergence of a global economy that is becoming more tightly integrated on a day-by-day basis
- The realization that the era of cheap energy has ended
- A growing scarcity of critical resources such as water, and more specifically, water in places that are significant population centers or agricultural regions
- The growing acceptance of global warming and its root causes

A vocabulary and set of catch words to support this concern for the environment has evolved, for example:

- Green
- Recycling
- Eco-friendly
- Carbon footprint
- Water footprint
- Organic
- BST-free (used to increase milk production in dairy cows)
- Freon-free (Freon is a gas that depletes the earth's ozone layer)
- Greenhouse gas (GHG)
- Gray water (polluted water)
- Smog (air pollution)
- Corporate social responsibility (CSR)
- Sustainability
- The triple bottom line (TBL): people (human capital), planet, profit

Nations have independently and collaboratively created standards to protect the environment, as well as the United Nation's sponsored initiatives. In support of these standards, methodologies and techniques have been developed, with the latest focus on the end-to-end supply chain as opposed to the narrow view of a stand-alone factory or enterprise.

This extended focus is a reflection of the maturity of the sustainability movement. For some individuals and corporations sustainability and social responsibility have become tightly linked. Current sustainability thinking has built on the initial areas of concern, such as reduction or elimination of packaging waste, and carbon management, to include water management. Sustainability, when linked to social responsibility, further adds fair trade, conflict materials, and human rights. Think of sustainability as a steel cable, such as that holding up the Golden Gate Bridge; both have multiple strands and are engineered for a collective purpose.

The book *Crossing the Chasm*, by Geoffrey A. Moore, segments individuals or groups into three categories when dealing with new technology or significant change. In the first category are early adopters, followed by mainstream adopters, and, finally, laggards.

Corporations that are early adopters or pioneers tend to be thought leaders within their industries and are commercially successful. Corporations that committed to the sustainability operating philosophy in its infancy have often garnered substantial rewards:

- A positive perception of them by the consuming public, a public that will even pay more for products and services from a company that is socially responsible
- Significant productivity improvements
- A reduction in raw material costs
- The ability to create innovative products that will replace existing carbon- or water-intensive products

At this point in time, sustainability is starting to be recognized by those organizations that are well managed but mainstream (versus early adopters). Some of recognition might be connected to the influence of the GHG Protocol, and Carbon Disclosure Project, and major institutional investors, who in essence are requiring a report card on sustainability actions before investing. These powerful influencers are examined in this chapter. Perhaps the mainstream classification needs to be segmented into early mainstream, mid-mainstream, and late mainstream.

Early adopters are now starting to look not just within the four walls of their enterprise, but at end-to-end supply chain sustainability. This end-to-end view recognizes that a carbon and water footprint is stored in all raw materials, and when aggregated together, provides a more accurate picture of the sustainability of a product.

A very sophisticated view of a product's sustainability includes the post-production logistics activities, the product's use by the consumer, and its disposal in a landfill, or hopefully recycling center. The methodology to gather the needed data for the carbon footprint (GHG Protocol Scope 3) from across the entire supply chain and report it has been recently developed, as has a comprehensive approach to managing water both locally and across the supply chain (Water Footprint Network). Both the GHG and water footprint methodologies are examined in this chapter.

The Supply Chain Operations Reference (SCOR) model has evolved over a fifteen-year period, and has been proven to work in any industry. It is a supply chain modeling approach that can capture the specifics of an existing end-to-end supply chain. Any organization committed to sustainability would benefit from using the SCOR Framework (developed by the Supply Chain Council, a nonprofit association), and then planning a more efficient future state. SCOR's basic purpose is to make a supply chain rational and efficient as well as to manage risk. Early on, SCOR's developers recognized that recycling was important and added it as a step called return. The process flow for each supply chain participant then evolved into plan, source, make, deliver, and return. Any organization that models its existing supply chain would be well served mastering and using SCOR as the foundation and frame of reference for its use of the GHG (carbon) and Water Footprint Network methodologies.

The rest of this chapter briefly examines:

- Recycling
- The SCOR Framework
- The Lean system's contribution to sustainability
- The Water Footprint Network methodology
- Approaches to managing the corporate carbon footprint
- The UN Compact and fair trade sources of supply

RECYCLING

Most organizations start their sustainability initiative with workplace cleanup and reorganization, and recycling. For food manufacturers, most of the production areas are clean and organized as mandated by current good manufacturing practices (cGMP), but there is always opportunity for improvement, and the Lean technique 6S is the tool of choice. Recycling and 6S have some things in common: Both are easy starting points involving minimum risk that generate excitement within the workforce, and are easy to see. Recycling can have three distinct focuses:

1. Analyzing what is being thrown away and salvaging what could and should be recycled. The simple approach of emptying external dumpsters and wastebaskets within administrative areas, laying everything out for visual examination, and brainstorming as to what should be recycled, and if it is internally generated packaging, determining if it is a candidate for redesign to minimize waste. Generally there is a significant amount of heretofore waste that can be recycled. One often overlooked consideration is the energy required for certain items to be recycled. This information is available on the Internet.
2. Designing out the need for or possibly redesigning internal packaging and consumable materials. This is often the responsibility of industrial or manufacturing engineers, but workers are a rich source of usable ideas. Knowing both the energy and water required to produce each item is important, and the information to roughly estimate resource usage is available from proxy data sources, such as the USDA. Did you know that one sheet of copier paper requires the consumption through its end-to-end supply chain of 2.5 gallons of water?
3. Getting your suppliers to reduce packaging and use recyclable materials. This will require that the customer (you) coach your suppliers, and if they are important enough to your day-to-day operations, insist they package in recyclable materials. At first their costs might be higher, but over an extended period of time, less so. Sharing with them the carbon footprint and water footprint data on packaging is important. There is another distinct possibility: One or more of your suppliers may be more advanced in their sustainability knowledge. Question them, and if they are, tap their expertise.

A cautionary note: There are trade-offs with recycling some materials, as the recycling processes needed for them are energy intensive. The information needed to make these trade-offs is readily available on the Internet.

In summary, recycling and the awareness that carbon and water can be designed out of materials are easy first steps that create a positive organizational mindset—plus it has a decent ROI.

THE SCOR FRAMEWORK

SCOR has been fully discussed in earlier chapters. In summary, it is a methodology to model an existing end-to-end supply chain as well as each organization (node) operating in it. SCOR has many advantages:

- It has been developed by a mix of industry thought leaders on a collaborative basis.
- It works for all industry segments, i.e., food, electronics, ship building, etc.
- It has evolved over a fifteen-year period and has been refined based on the experiences of corporations using the methodology.
- It is successfully used by many organizations.
- It is constantly being enhanced and updated; i.e., supply chain risk was added to the methodology approximately five years ago.
- It is maintained by a nonprofit association.
- It has become codified into a teachable form that is easily accessed.
- It has become a de facto standard.

At the highest level, SCOR uses five major processes—plan, source, make, deliver, and return—to model an organizational participant in a supply chain. Figure 8.1 depicts the high-level modeling. Each of the five process steps can cascade down into progressively greater levels of detail, with up to four additional lower levels. The fourth and lowest level is highly customizable.

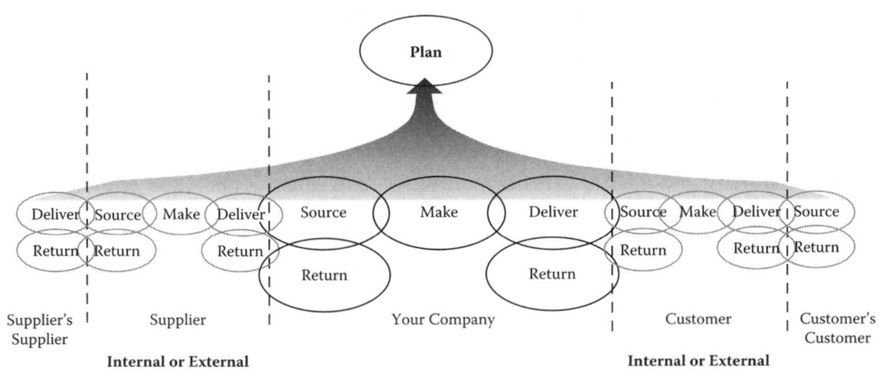

FIGURE 8.1
A high level conceptual view of the SCOR model of Plan-Source-Make-Deliver-Return. (From Supply Chain Council. With permission.)

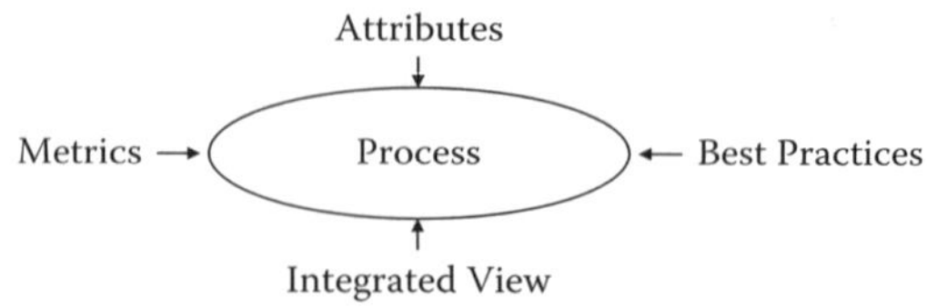

FIGURE 8.2
What is applied to each of the process steps of Plan-Source-Make-Deliver-Return in SCOR.

SCOR supports three operating styles and market supply philosophies: make to stock, make to order, and engineer to order. In addition, all five processes (plan, source, make, deliver, and return) have attributes attached to them as well as performance metrics. The third and final attachment is best operating practices. Figure 8.2 is a simplified visualization of this integration. Modeling an existing supply chain is the first step in the methodology. The ultimate goal is the improved performance of a supply chain, or its alignment with corporate strategy. "If you don't know you're in Detroit, you won't know how to get to Atlanta."

This modeling requires the development of the targeted future-state supply chain. The developer of SCOR has put in place a method to determine what should be improved, and has arranged confidential benchmarking services. This service compares companies within an industry type, i.e., dairy products, flavors, poultry, etc., but also factors in additional considerations, such as company size and geography. It also segments the results into that industry's best performing, average performing, and underperforming, so there is a sense of where to focus and how big of an improvement is needed. This benchmarking information, in combination with the current-state supply chain map, enablers, and best practices (defined in the methodology), is the information needed to develop the future state, strategy, and tactics to achieve that improvement.

Since many companies have multiple supply chains, the focus should be on those that best support the organization's overall competitive strategy.

Thanks to SCOR the ability to understand, visualize, and improve an end-to-end supply chain is possible. It lays the groundwork for sustainability schemes to plug in to, such as carbon and water management.

THE LEAN SYSTEM'S CONTRIBUTION TO SUSTAINABILITY

For many, Lean thinking is associated with the elimination of waste in all forms, but with special emphasis on removing the following excess:

- Time
- Inventory
- Labor
- Material handling
- Factory and office space
- Administrative processes
- Data

Recently, the elimination of excessive energy use, with many corporations embracing the Lean management operating system, has become an important objective. Traditionally efforts within a factory to minimize electrical energy use have focused on the devices used to process product. Much good has been accomplished with this focus, and with continuing innovation there is undoubtedly more progress possible.

The good news? Applying continuous improvement techniques complements this traditional focus. These techniques are centered on how work is performed by using energy-efficient equipment, and in some cases eliminating the need for any resource-consuming machine. One of the most surprising insights gained from Lean projects has come from the reduction of support energy, and often this has a greater reduction in energy waste than the direct manufacturing equipment itself. Air compressors, lighting, cooling towers, and backup systems have been seen as vital to running manufacturing operations, and with limited control by plant managers.

Collaboration with an Electric Energy Provider

> The U.S. DOE [Department of Energy] has stated that the internal rate of return is above 20% for energy reduction projects.
>
> **—Gerald Church, Energy and Environmental Manager, California Manufacturing Technology Consulting (CMTC)**

One example of a program to reduce the carbon footprint through the application of continuous improvement (CI) techniques was the initiative by a major electric utility company that wanted to stabilize electricity consumption within its manufacturing customer base.

The energy provider served a geographic area that had experienced considerable population growth and energy consumption, and additional growth was anticipated. Building a new power plant is expensive, and requires considerable environmental reviews and time. It's a mega-project. Per capita consumption within the residential customer base was low, thanks to the utility provider offering incentives to use energy-efficient appliances, and retrofitting homes to conserve energy.

The next biggest opportunity to reduce energy consumption was within the manufacturing industry. The electric energy provider did have existing programs to incent manufacturers to use energy-efficient machinery and facilities, but it wanted to go beyond just focusing on individual devices to reduce energy consumption and look at opportunities to reduce the carbon footprint for entire processes. The electric utility was convinced that CI techniques could not only eliminate the traditional wastes of labor, inventory, and time, but also reduce energy consumption. It found another large and neighboring energy utility that also shared this conviction, and this collaboration was a logical progression on both organizations' previous efforts to improve customer energy conservation and that of the U.S. DOE. Both wanted to put together a pilot program with a local expert in the application of the Lean system and change management strategies to support behavior modification in furthering resource sustainability.

It looked to a nonprofit consultancy that specialized in helping companies adopt CI techniques. This consultancy is a member of a network of consultancies within a program supported by the U.S. government; its purpose is to help manufacturers of all sizes improve their operating efficiencies and competitiveness. A joint plan was developed and rolled out. Manufacturers with high energy consumption were targeted if they had demonstrated through their internal investment actions a commitment to energy usage reduction. In the geographic area both the energy utilities serviced, many manufacturers volunteered to support rolling blackouts during summer months, when there were excessive heat episodes. Using less energy was a win-win for every stakeholder in the community. These progressive manufacturing organizations recognized the following truth:

> A 2004 DOE report estimated that perhaps 70% of the opportunity to improve U.S. manufacturing plant energy efficiency was in improving plant equipment energy efficiency applications and about 30% was in improving plant energy management practices (DOE 2004).
>
> Regardless of the percentage, the point is that in addition to potentially improving equipment energy efficiency, there is a significant opportunity to improve U.S. manufacturing plant energy management practices.
>
> **—Gerald Church, Energy and Environmental Manager, California Manufacturing Technology Consulting (CMTC)**

The plan stressed the use of the value stream mapping (VMS) technique with one innovative addition, the addition of the energy used within each major "node" or process step within an overall process. Value stream mapping was discussed in an earlier chapter. This enhanced technique is known as value and energy stream mapping (VeSM). Based on the road map a VeSM provided, and the philosophy of the Pareto rule, a rule that states that eighty percent of value (or waste) is concentrated in twenty percent of activities, kaizen events were scheduled and other CI techniques were used selectively. Another area of innovation was the development of energy usage measurement methods.

Not all manufacturers would have the same configuration of equipment types, nor are at the same point in their efforts, nor have the same expertise in reducing energy consumption. To work with this wide range of possibilities a number of techniques were developed to accurately isolate points of consumption that could be used in various combinations.

A plan without measurable objectives is a wasteful exercise, so a goal of a ten percent reduction in energy use was established. This ten percent was above and beyond the benefits of using newer energy-efficient equipment; it focused on:

- Improved methods of performing work
- Organizing to do work
- Streamlining and simplifying workflows

All with the goal of reducing energy consumption *and* increasing capacity.

This approach was very attractive in that it had an economic multiplier—qualified manufacturing process improvements typically yield labor or materials savings, plus equivalent plant capacity dollar value savings, that are several times annual energy dollar savings.

Another attraction was that improvements were generated using the Lean system's continuous improvement (CI) techniques. These techniques generally require low-cost expenditures for supplies, tools, and the like, and avoid the requirements associated with large capital expenditures. This removed a significant barrier in many plants to getting started with energy savings initiatives.

Some Specifics

The areas of potential energy reduction surfaced in a VeSM became candidates for a kaizen event. The focus during a kaizen event was on three major possibilities:

1. Scrap reduction through the application of Six Sigma. Less scrap to yield the desired amount of product means less energy is used, if the equipment used in a process is used only when producing a product batch, and idle in between batches. The description used for this is direct energy use equipment.
2. Increasing the velocity of product moving through a process or set of processes. This shortens the time a product is in a production cycle. This translates into less use of generalized plant equipment that is needed, irrespective of which item is being produced. This equipment is really in the overhead category. An example: If eight hours of the production time required can be reduced by X hours, there is less need to heat, or cool, or provide illumination in a factory for eight hours. This energy use was titled indirect equipment energy use.
3. Reducing energy consumption for products processed through specific individual machines that are single purposed. This is a variant of the direct energy use equipment.

There were many other measures and calculations developed that the reader is encouraged to examine in the future.

A Mini-Case Study

A food products manufacturer conducted two kaizen events with the following improvement objectives:

- Reduce equipment changeover time

- Increase equipment run-time through:
    1. Better equipment maintenance
    2. Faster equipment repair/recovery
    3. Better start-up procedure checks to reduce the risk of an unexpected line shutdown, etc.

The combined results from the kaizen events increased actual equipment run-time percent of shift hours worked (actual line capacity) from an average of just under fifty percent to about eighty percent, for a sixty percent increase of available plant capacity per work shift. These projects saved about 240,000 kWh/year. The sixty percent gain in plant capacity enabled the company to significantly increase plant production and sales for a very large financial company benefit.

There is a link to another mini-case study in the Appendix under the subheading "Sustainability."

Others Who Have a Vested Interest in Energy Usage Reduction

Traditional methods of producing energy generate an unwanted by-product: pollution. There is no doubt that energy producers have made significant progress over the last few decades in reducing pollution. Also, nuclear generated power doesn't seem to be acceptable to the public; it has risks associated with it that for many offset its advantages. Yes, "green" energy is on the horizon, but change takes time, and "now is now," and pollution is damaging. That is why the U.S. Environmental Protection Agency (EPA) has shown interest in the reduction of energy use through increased manufacturing efficiencies that in turn reduce pollution, yet still support economic considerations.

While the examples used so far in the chapter are U.S. centric, the desire to reduce energy consumption is global, and knows no borders. A good example is the European Union (EU). The EU is industrialized and has a large food manufacturing industry. Collectively it has not been energy rich, so the EU shares the same concerns as its global partners about energy conservation. Also, the Lean system is universal and can be embraced by any progressive food company, irrespective of its geographical location; it just works!

Mini-Summary

- There is growing concern about the traditional forms of energy generation, its consumption, and how excessive usage negatively impacts our ecosystem. Also, the definition of *normal* versus *excessive* has changed! What was normal twenty years ago is now considered excessive.

  This, in combination with the realization that there will be resource shortages in the future, energy being one of many resources in that category, has prompted major governmental agencies such as the U.S. FDA to encourage the food industry to be proactive in reducing energy consumption.

- Strategies and tactics to minimize energy consumption have started to mature, and are rapidly evolving. The next ten years will be explosive in their relevance and application compared to the previous ten years.
- Historically the first step or leg for many companies on the sustainability journey is focused on recycling.
- Recycling is great for increasing organizational awareness because it's very visual and collaborative.

  It often moves in stages; progressing from sorting through dumpsters and trash containers, with a focus on recycling waste that previously became dangerous and expensive landfill. The second stage is centered on reducing or eliminating packaging of products shipped to customers, followed by the packaging of goods received from suppliers. While this might sound simplistic, it can be really invigorating to an organization and sets the stage for other sustainability efforts.

Many thought leaders in sustainability have looked to Lean continuous improvement techniques to reduce energy consumption, increase operational efficiencies, and the effective and waste-free use of organizational assets.

Sustainability is now starting to move from an enterprise perspective to that of an end-to-end supply chain. Yes, the enterprise is important, but it is part of a network called the supply chain that collaboratively delivers benefit. The SCOR Framework from the Supply Chain Council codifies and makes possible the analysis of an entire supply chain. This sets down a foundation that sustainability analysis tools such as the Water Footprint Network and the Greenhouse Gas (GHG) Protocol can overlay on, and combines traditional and understood economic reward (SCOR) with the often ignored and misunderstood rewards of sustainability.

THE WATER FOOTPRINT NETWORK METHODOLOGY

Introduction

Energy is important to both the food and nutritional supplement manufacturer. But water is even more important. And its availability historically has, at least by the public, been taken for granted.

Nature has a mind of its own: It puts water where it wants, not necessarily where we want it to be; drought in one area, flood in another. In the food supply chain, water is required for the production of almost all raw materials. Agricultural products are water intensive with a few exceptions.

There is a growing need to reduce water intensity at every process step in the supply chain, with the new viewpoint that there is water embedded in energy and energy embedded in water, which is giving rise to integrated energy and water savings programs at organizations that have historically been industry thought leaders.

The raising or production of meat is also water intensive, not only in a direct consumption sense, but also in the agricultural products (feed) the animal in the preprocessing stage consumes. This indirect consumption of water is called virtual water, water used in the production of the livestock's food. The concept of embedded or virtual water is relatively new, as is the accompanying concept of cumulative inheritance as a food product progresses through its supply chain as well as its consumption of all raw materials and their associated direct and virtual water.

Both of these new understandings are part of an overall water usage assessment, accounting, future planning, and management methodology from the Water Footprint Network. Figure 8.3 depicts the interaction of direct and indirect water and its progression "downstream" in the supply chain. Notice that step 2 carries the direct water from step 1 forward as

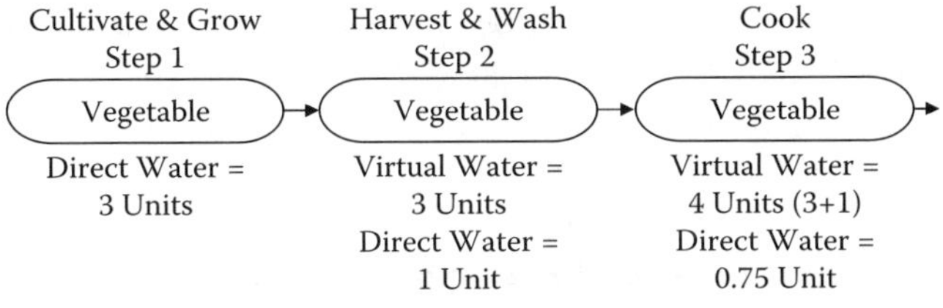

FIGURE 8.3
Moving downstream in the supply chain. Water content: direct (processing) and virtual (embedded/indirect).

100 Grams of Chocolate
2,400 Liters of Water =
634 US Gallons

2.2 Pounds of Refined Sugar
1,500 Liters of Water =
396 US Gallons

1 Hamburger
2,400 Liters of Water =
634 US Gallons

FIGURE 8.4
Examples of the total water content (virtual+direct) for food products. (From Arjen Hoekstra, Water Footprint Network. With permission.)

virtual water, and step 3 carries the direct and indirect water from step 2 forward as virtual water plus its direct water usage.

When the consumption of water by all food ingredients in a recipe collectively moves downstream (moving forward), through each of their processing steps in a supply chain, and they come together in a production batch for the final product recipe, the cumulative total amount of water consumed is much greater than one would assume; it can be an eye-opening discovery! Figure 8.4 highlights a few examples.

Twenty years ago if someone proposed that an organization could accurately determine the water usage in its supply chains, including both direct and indirect water, that individual would be considered a futurist and not grounded in reality. But today, it is possible with the convergence of methodologies that are complementary. One of them is the SCOR Framework that is focused on supply chain efficiencies, flexibility, and responsiveness. SCOR has a traditional economic focus, but a supply chain modeled by it can provide a foundation for methodologies dedicated to sustainability across the supply chain, such as the Water Footprint Network and GNG Scope 3 (carbon) methodologies, both of which have a science orientation. There are other beneficiaries of the SCOR foundation that are humanitarian in nature, such as fair trade and conflict minerals.

More Information on the Water Footprint Methodology

Thus far only the ability to calculate both direct and virtual water (inherited from previous processing) has been discussed. There is much more to this comprehensive methodology:

- It is truly consumption based, in contrast to the traditional and misleading water consumption accounting approach(es).
- It recognizes three different types of water: green (rainwater), blue (subterranean and ground), and gray (polluted).
- Water inflows and outflows are incorporated into the methodology. A blue water source isn't static.
- It provides insights into the management of gray (polluted) water.
- It has a full library of workable standard calculations to support the goals of the methodology.
- It has a geographical perspective that connects the local, regional, and global geographic views, which facilitates an accurate assessment of a supply chain's water consumption, and supports the development of a desirable future-state supply chain.
- It recognizes seasonality and ecological "hot spots" within the geography where any contributor to a supply chain resides.
- It supports multiple types of assessments: global, national, regional, local, industrial, specific food processor/manufacturer, and individual consumer.

Traditional methods of calculating water consumption are misleading. Most are focused on only blue water directly consumed locally. Blue water is the water in aquifers, rivers, lakes, etc., that with further processing can be made drinkable. Figure 8.5 graphically depicts this perspective.

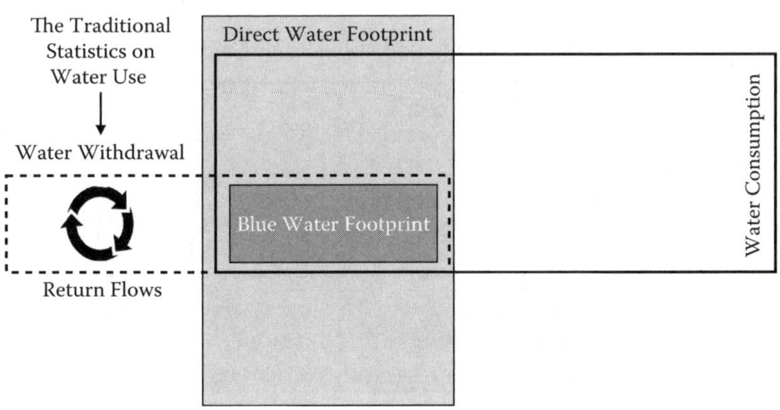

FIGURE 8.5
The traditional and misleading method of calculating water consumption. (From Arjen Hoekstra, Water Footprint Network. With permission.)

The Water Footprint Network methodology recognizes the full spectrum of water types:

- Blue water
- Green water—rainwater that can directly irrigate agricultural crops, etc.
- Gray water—polluted water from various sources that must be diluted with blue water to achieve an acceptable biological state before being discharged into a water ecosystem

Figure 8.6 depicts this comprehensive view.

What also contributes to the Water Footprint Network methodology's accuracy as a water consumption assessment tool is the addition of the concept of embedded or virtual water that is stored in the products and services consumed locally.

Figure 8.7 shows the supply chain for the production of a cotton shirt and how much virtual water is passed on to subsequent processors, who then use direct water to complete their value-adding contribution. For those who have used an Enterprise Resource Planning (ERP) system to calculate total product costs using the cost roll-up facility, the underlying logic will be familiar.

Water Inflows and Outflows—Water Isn't Static

The Water Footprint Network's methodology (aka the global water footprint standard) not only puts in place the critically important factors

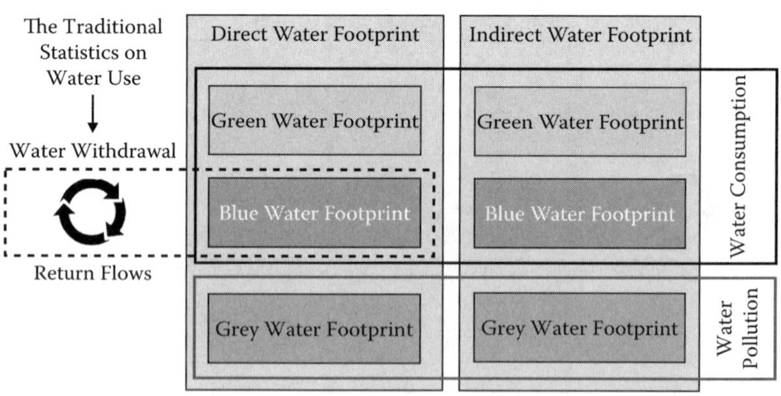

FIGURE 8.6
A high level comprehensive of water types. (From Arjen Hoekstra, Water Footprint Network. With permission.)

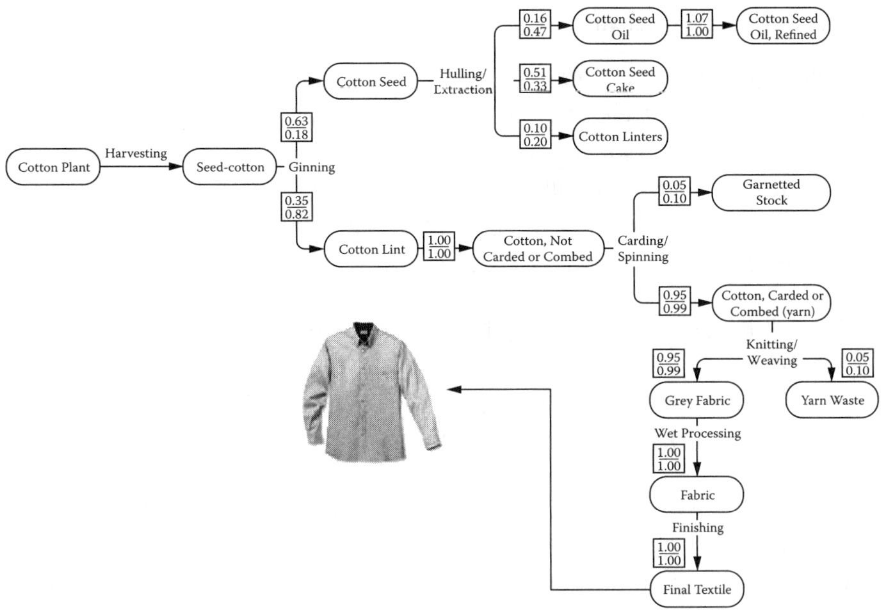

FIGURE 8.7
Cumulating both the indirect (virtual/embedded) and direct water as it progresses downstream in the supply chain. (From Hoekstra and Chapagain, Water Footprint Network, 2008. With permission.)

required to accurately determine true water usage, factors such as water type (green, blue, or gray), and the presence of virtual water, or water embedded in products sourced outside of the local consuming entity's geography, but also has a deep scientific understanding of the "rhythms" of a water catchment and its ecological health. Figure 8.8 shows the integration and cycling of water flows.

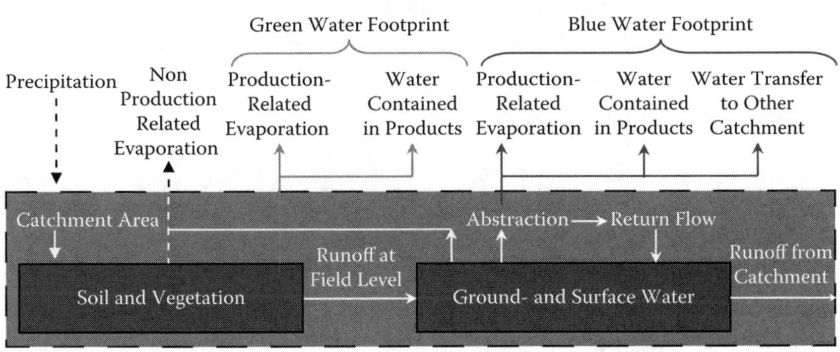

FIGURE 8.8
A deeper look at the behaviors of water. (From Arjen Hoekstra, Water Footprint Network. With permission.)

Gray Water Pollution

The inclusion of the gray water concept and the science it uses is very important in managing water because of the large-scale agricultural and industrialized food processing systems in use today. Gray water is a by-product of chemical and thermal pollution generated in a processing step of a product. It is one of the three footprints in the water footprint methodology.

The definition of gray water is as follows:

> The volume of water that is required to assimilate waste, quantified as the volume of water needed to dilute pollutants to such an extent that the quality of the ambient water remain above agreed water quality standards.
>
> **—Arjen Y. Hoekstra et al., *The Water Footprint Assessment Manual***

The ambient water condition is central to gray water calculations within the methodology, because it is the condition of the receiving freshwater body that determines the treatment and amount of gray water discharged into it.

Ambient water quality standards are a specific category of standard, and different from other quality standards, such as drinking, irrigation, and effluent water. Ambient water quality standards may vary from place to place because the natural concentration (or profile) of the receiving freshwater catchment may vary. Each ambient water catchment is treated as unique within the methodology because of its ability to assimilate a certain pollution load. The pollution or critical load considers the natural concentration and maximum supportable concentration of the pollutant being managed in the receiving water catchment. At the critical load, the gray water footprint will be equal to the available water flow, which is then required in full to dilute the chemicals down to acceptable concentrations. Since there can be more than one polluting chemical the current recommendation is to concentrate on the most damaging and critical.

A few other interesting points:

- It might be appropriate to have two gray water footprints, one for groundwater and a different one for surface water.
- Groundwater quality standards are often based on requirements for drinking water, and surface waters are typically determined by ecological considerations.
- Evaporation can cause a specific form of pollution in that it will concentrate a chemical after the water evaporates. An example is the high salt concentrations in drainage water from irrigated fields.

- Wastewater treatment reduces the gray water footprint, but in all probability doesn't fully eliminate it.
- Recycled water may have a limited number of uses and have to be eventually discharged.

In summary, gray water footprints are measured based on the (human-induced) contaminate loads that enter into freshwater bodies. How to measure a gray water discharge and how to remediate it is clearly defined in the water footprint methodology. Treatment of gray water to reduce contaminates prior to release reduces that load. The goal should be recycling of gray water and, if possible, elimination.

Workable Standard Calculations

A methodology isn't complete or usable until it has the following:

- A dictionary of standard terms
- A list of standard symbols used in calculations
- Usable calculations to support each phase of the methodology
- Standard courseware to teach others about the methodology
- Flexibility in its application
- Forums to evolve and improve the methodology
- Acceptance by credible authorities

The water footprint methodology satisfies all of the above criteria. In the Appendix the reader will find definitions used in this chapter, but they are just a partial list of the words developed or incorporated into the methodology. Also in the Appendix is Recommended Reading and the URL for the Water Footprint Network, and on their website is the list of involved or sponsoring entities, ranging from government agencies to thought-leading associations and corporations.

Since the operations and supply-chain management practitioner is calculation oriented, a simple calculation for the gray water footprint is listed in Figure 8.9. Most calculations are more complex.

The highest level of analysis is at the footprint level, but there are many subordinate calculations to support it. A partial list of specific footprint calculations available is in Table 8.1. The units of measurement for all footprints are volume and time.

$$WF_{PROC,GRAY} = \frac{L}{C_{MAX}\ C_{NAT}} \quad \text{[Volume/Time]}$$

Symbol Definitions

WF

PROC, GRAY

Volume/time

Gray Water Footprint of a Process

L

Mass/time

Load of a pollutant

C

MAX

Mass/volume

Maximum acceptable concentration of a chemical in a receiving water body

C

NAT

Mass/volume

Natural concentration of a chemical in the receiving water body

FIGURE 8.9
An example of the calculations within the Water Footprint Network methodology. (From Arjen Hoekstra, Water Footprint Network. With permission.)

TABLE 8.1

A Partial List of the Symbols and Their Meaning within the Water Footprint Network Methodology

Footprint Symbol	Explanation
WF_{area}	Water footprint within a geographically delineated area
$WF_{area,\ nat}$	Water footprint within a nation
WF_{bus}	Water footprint within a business
$WF_{bus,\ oper}$	Operational water footprint for a business
$WF_{bus,\ sup}$	Supply chain water footprint of a business
WF_{cons}	Water footprint of a consumer, direct + virtual (embedded/indirect/imported) water
$WF_{cons,\ direct}$	Water footprint of a consumer, direct usage, no virtual
$WF_{cons,\ indir}$	Water footprint of a consumer, indirect (virtual/embedded/imported)
$WF_{cons,\ nat}$	Water footprint within a nation
WF_{proc}	Water footprint of a process (all types)
$WF_{proc,\ blue}$	Blue water footprint of a process
$WF_{proc,\ gray}$	Gray water footprint of a process
WF_{prod}	Water footprint of a product (all)

Again, this is just a partial list of water footprints, all of which have analytical and calculating logic to determine specific values in just about any situation. The water footprint methodology has a "rich" and comprehensive toolset to help any corporation, government, or person get a status on real water usage, and put in place a plan that has economic, ecological, and social merit.

Global Geographic Perspective

During the early days of the sustainability movement the focus was on the direct water usage of a single entity. In the context of the food industry, it was the farm or food processor.

The recognition of gray water was considered only a local compliance issue, not a water usage issue of scarce blue water. Business trade was local or national. Concern for the ecology was noble, but had little economic implications.

Things have changed over the last twenty years. Today we operate in a tightly interconnected and fluid global trading community. Concern for the ecology is not just an exercise in idealism, but a growing recognition that it has an economic upside, a really attractive return on investment. The meshing of ecology and commerce is recognition that there will be significant resource shortages in the future. Shortages create a seller's market that in turn creates pressures for efficient use of them. Blue water and, in some situations, green water are among the most significant resource shortages today.

Ironically, we have much less water in places where we have created infrastructure, in cities and businesses, and in some cases extreme shortages. This is in contrast to carbon, in which we also have shortages (oil), and its undesirable by-product, carbon, in abundance!

Fortunately, the focus has now moved to a global view, building on top of the local view, and we are now trying to manage end-to-end supply chains for not only economic reasons (the SCOR methodology) and concerns about global warming (Greenhouse Gas (GHG) Protocol Scope 3) but also water, with the Water Footprint Network methodology. Figure 8.10, which is similar to the SCOR visual global mapping of a supply chain, provides the correct perspective.

Seasonality and Ecological Hot Spots

Global supply chains mean that different supply chain partners can experience seasonal differences between them. Even within a medium-sized geography,

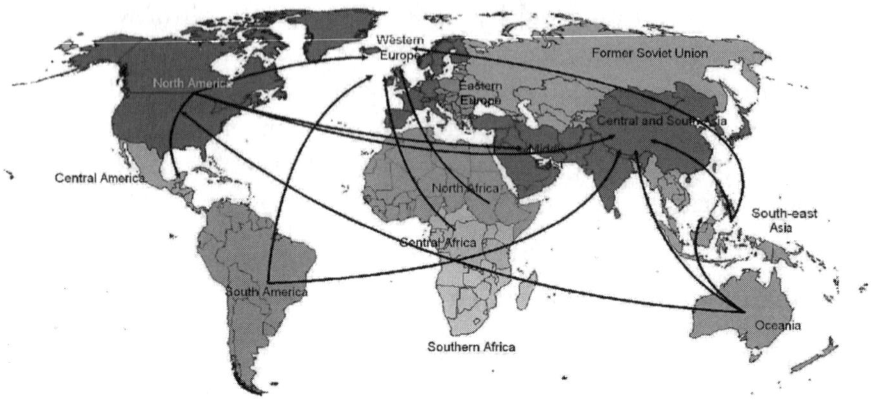

FIGURE 8.10
Virtual water was a hidden global problem and masked the blue water problem until now. (From Arjen Hoekstra, Water Footprint Network. With permission.)

for example, Southern California, there can be multiple micro-climates. In the case of Southern California there are five—coastal, inland coastal, foothills, mountain, and desert—all within a hundred-mile (160.9 km) radius with daytime temperatures ranging from 61 to 120°F (15.56–48.9°C).

Factor in rainy and dry seasons for any region, and that translates into variable natural water replenishment for many of those participating in a supply chain. That means an upstream (supply chain jargon) supplier, growing, or processing product during the dry season (or drought) might be stressing the ecosystem of its local water supply. This is not good and can have a devastating economic impact on a local community or region. An unfortunate but sobering example is that of the Aral Sea region, where cotton was grown and used up most of the water resources.

The good news? The Water Footprint Network methodology emphasizes measuring and monitoring the eco-health of a water catchment on a regular basis, and provides the necessary body of knowledge to do so. Figure 8.11 highlights, based on the status of a catchment, its danger zone.

Multiple Types of Assessments

The obligation for sustainability is shared by every government, business entity, and individual. Unfortunately awareness of this obligation has grown unevenly within each ecosystem stakeholder group. In some areas it has been easy. An example is recycling; it is easy because it is tangible and visual. The next logical step in the recycling movement is to demand

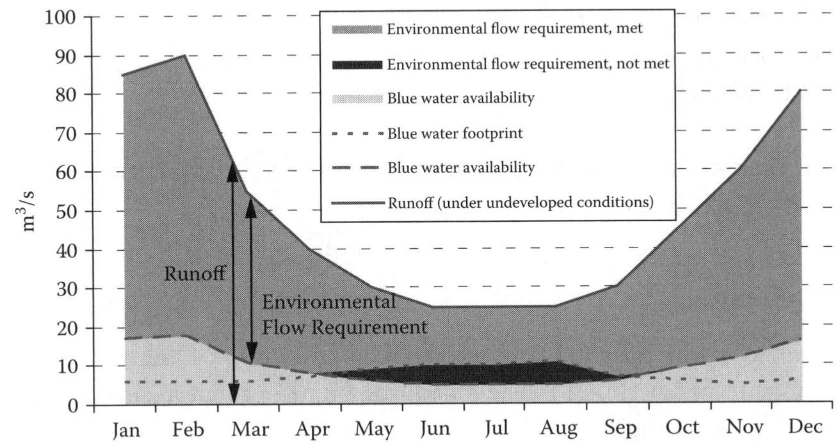

FIGURE 8.11
Monitoring the eco-health of a water catchment. (From Arjen Hoekstra, Water Footprint Network. With permission.)

minimal but safe packaging by all stakeholders. The growing awareness that there is a significant sustainability–profit and loss (P&L) relationship has been acted on by thought leaders in business and industry. There are still many laggards, but they will eventually follow the leaders.

Another contributor to the slow awareness has been the parochial or local view of the consumption of resources. With the rapid development of a truly dynamic global economy, the local view is no longer adequate. We now think in terms of a supply chain, not just of a factory, which is a major move forward and compatible with the way we do business today.

The biggest change is the availability of methodologies that allow for the analysis and designing of supply chains using the SCOR Framework, GHG Scope 3 (covered below), and Water Footprint Network methodology, thereby minimizing the consumption of critical resources, money, water, and carbon.

The Water Footprint Network methodology stands out as being easily used by multiple stakeholders:

- The individual (consumer)
- The business unit (grower, food processor, transporter, supply chain participant)
- A process owner (internal to a business)
- The supply chain orchestrator (closest to the ultimate consumer)
- National governments

- Regional governments
- Local governments
- Local water districts

No excuses. The tools to take action are there for all water usage stakeholders and just need to be applied.

Mini-Summary

A picture is worth a thousand words (see Figure 8.12). How are food and beverage industry thought leaders addressing water management and the tools they are using?

> Unilever's Sustainable Living Plan includes a target to halve the environmental footprint of our product portfolio across the life cycle. Water is one of our key metrics and we support efforts to standardise methodology and improve access to scientifically robust and standardized data. The Global Water Footprint Standard is an excellent resource and fundamental to us achieving our targets.
>
> **—Dr. Donna Jeffries, Sustainability Manager, Unilever**

> The Water Footprint Network has since its establishment provided a global, multi-stakeholder platform to advance the dialog around the quickly-emerging topic of water footprint. At PepsiCo, we believe that there is significant value in understanding the components of our water footprint from the perspective of local impact. In turn, understanding and addressing this local impact is a critical element in our commitment to respect water as a fundamental human right.
>
> **—Dr. Liese Dallbauman, Senior Manager of Water Stewardship, PepsiCo**

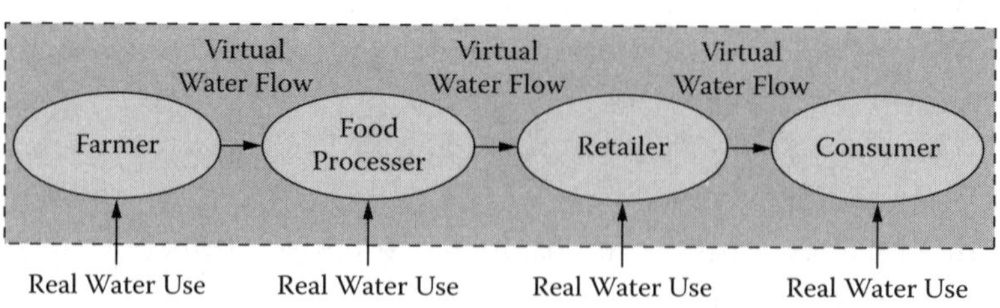

FIGURE 8.12
A simplified view of how virtual (embedded) and direct water accumulates. (From Arjen Hoekstra, Water Footprint Network. With permission.)

APPROACHES TO MANAGING THE CORPORATE CARBON FOOTPRINT

There are many nongovernmental organizations (NGOs), national governments, global intergovernmental forums, and nonprofit organizations involved in creating awareness of the current carbon footprint problem, with a strong thread of collaboration between them. Not only do they collectively have the goal of creating awareness, but they also develop methodologies to inventory existing carbon emissions within an enterprise, develop approaches to reduce emissions, and report current greenhouse gas (GHG) status and progress.

While the one might be inclined to think of heavy industrialized industries when discussing a carbon footprint, agriculture and food processing also have a significant footprint due to the materials used at the start or root of the supply chain, and the subsequent use of energy to heat, cool or freeze, store, and deliver (logistics) product.

The carbon footprint reduction movement in its various forms is less than twenty years old. It recognizes that significant carbon reduction not only is an ecological consideration and social obligation, but also has long-term economic consequences. This has been acknowledged by major thought-leading corporations, who have responded by sponsoring and supporting many carbon reduction forums and initiatives.

Reducing a carbon footprint isn't just only about eco-goodness, it's about a corporation's future profit or loss (P/L) potential. Carbon-based resources are no longer cheap, and will become progressively more expensive in the future.

Early efforts within the carbon reduction movement were centered on the four walls of the enterprise. Today it has been expanded to the entire supply chain. A convergence has occurred on understanding an end-to-end supply chain's current performance and improvement potential across resource dimensions:

- Managing the supply chain to achieve maximum traditionally calculated profit, with the maximum possible flexibility and minimum acceptable risk—the SCOR Framework
- Managing the end-to-end water supply chain to minimize all forms of water—blue, gray, green, direct use, and virtual (embedded/inherited)—with recognition of seasonal ecological hot spots

- Inventorying, understanding, and developing tactics to reduce a supply chain's carbon footprint—Greenhouse Gas Protocol (GHG Protocol)

Greenhouse Gas (GHG) Protocol

The *de facto* standard for inventorying, reporting, and GHG reduction techniques is the Greenhouse Gas Protocol. The GHG Protocol:

- Is a collaborative effort with the World Resources Institute (WRI), World Business Council for Sustainable Development (WBCSD), and International Organization for Standardization (ISO)
- Is a nongovernmental organization (NGO)
- Has been in existence for over a decade
- Is widely accepted by major thought-leading corporations
- Is a comprehensive GHG accounting tool with standard measurements, and a suite of calculation tools to understand, quantify, and manage GHG emissions
- Consists of four standards, with two foundational standards:
    1. Corporate accounting and reporting standard, a step-by-step guide for quantifying and reporting GHG emissions. This standard is focused on the figurative four walls of the enterprise.
    2. Project quantification standard, a guide for quantifying reductions from GHG emission mitigation efforts (projects). This also is focused on the figurative four walls of the enterprise.
- Is used by the Carbon Disclosure Project, a nonprofit organization whose inventory of corporate emission reports and analytical tools is used by major institutional investors

A simplified view of the collaborative nature of the GHG Protocol is shown in Figure 8.13. The GHG Protocol currently has four standards:

1. Corporate (establishing a baseline and reporting)
2. Project (quantification of reductions)
3. Product life cycle
4. Value chain (supply chain)

These are designed to be generic in nature, but are supplemented with special protocols to address the specific requirements of certain sectors, such as agriculture. There is also the convention of using the term *scope*

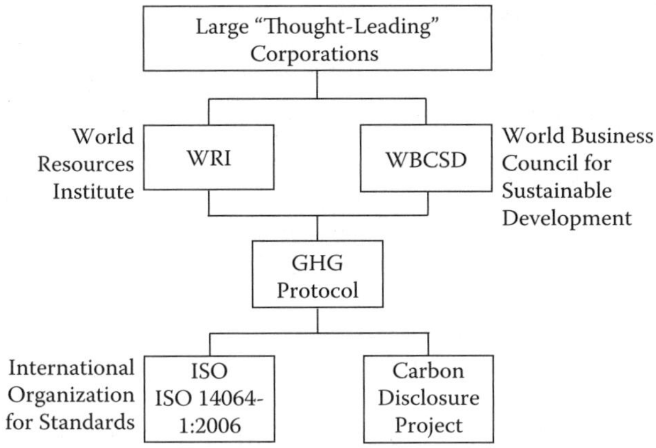

FIGURE 8.13
The GHG Protocol sponsors and collaborators.

to identify the boundary an organization has established to inventory its current carbon footprint, and report the status and progress in its carbon reduction journey. The nomenclature is:

- Scopc 1: The generation of carbon emissions within the reporting enterprise.
- Scope 2: The carbon emissions figuratively embedded in electricity purchased from external sources. Carbon generated during the creation of electricity.
- Scope 3: All the carbon emissions generated both upstream and downstream in an enterprise's supply chains, including purchased electricity (Scope 2). In the GHG Protocol it is referred to as the value chain. Scope 3 is the latest development in the evolution of the protocol.

Figure 8.14 highlights what each GHG Scope covers.

The first standard in the GHG Protocol (there are three others—project, product, and value chain) is titled the corporate accounting and reporting standard. This standard is where the compiling of the GHG inventory takes place. There is strong business logic for a corporation to commit to this standard, and the GHG Protocol guidance document highlights the following benefits, and the author has added supplementary comments:

- Manage GHG risks and identifying reduction opportunities. Carbon usage will become a significant cost consideration (burden) in the future.

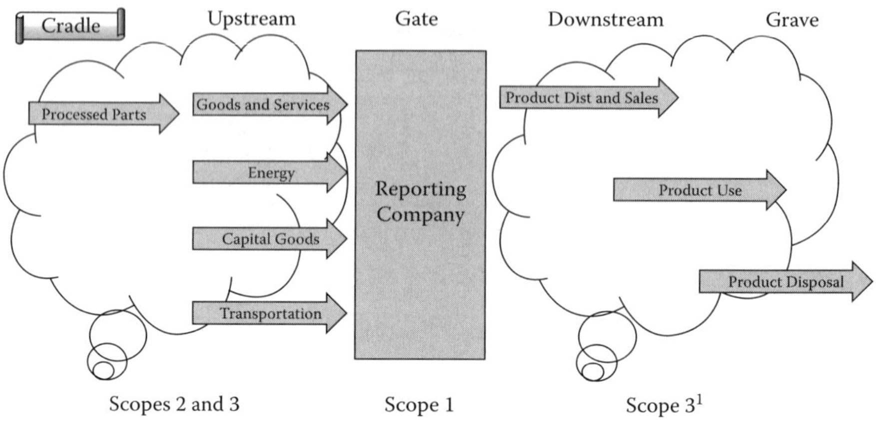

FIGURE 8.14
The GHG Scopes and what they cover in the end-to-end creation of products and services. (From Kuhn Associates. With permission.)

- Publically traded corporations can report their participation in voluntary GHG programs. This projects a positive image with both commercial and industrial consumers in a very competitive global environment, and with large institutional fund investors.
- Make mandated GHG reporting easier when required by governmental entities, or large customers, such as Wal-Mart, Cargill (United States), Miller Brewing (United States), Unilever, Starbucks (United States), etc.
- Recognize voluntary action. The idealistic consumer, both commercial and industrial, will always support the eco-sensitive organization, patronize it, and if reasonable, pay a higher price.

What gets measured gets managed.

The corporate accounting and reporting standard is the starting point and foundation for emission accounting and reporting, and is similar in intent and logic to the required financial reporting for publicly traded corporations.

This standard has been crafted with the following accounting and reporting principles:

1. Relevance: Contains the information on GHG activity within an inventory boundary (organizational unit) that stakeholders need to make informed decisions.
2. Completeness: The inventory of GHG sources is comprehensive, compiled, and meaningful in determining the carbon footprint within an inventory boundary.

3. Consistency: In the gathering, compiling, assessment, and publishing of GHG-related data within an inventory boundary, including assumptions and the known limitations of the techniques used.
4. Transparency: On the methods used to gather and compile GHG data within an inventory boundary, including assumptions and any known issues in estimating a GHG contributing process. The goal is a clear, factual, understandable, and meaningful report.
5. Accuracy: Data should be accurate and precise enough to be credible, and support decision making.

The corporate and reporting standard has three distinct groupings (consisting of nine activities or process steps) that the accounting and reporting principles are applied to: planning, managing, and reporting.

Planning:

- Identifying and choosing business goals and inventory design
- Setting organizational boundaries
- Setting operational boundaries

Managing:

- Identifying and calculating GHG emissions
- Tracking GHG emissions over time
- Managing inventory quality
- Accounting for inventory reductions

Reporting:

- Verifying GHG emissions
- Reporting GHG emissions

A Brief Synopsis of the Nine Activities or Process Steps

1. Identifying and Choosing Business Goals and Carbon Inventory Design

Effective management of the carbon footprint starts with the senior management team, because the footprint has within it significant threats and opportunities that have evolved and been recognized over the last few decades, and the threats will become more intense in the future. Carbon emissions are currently expensive, and will become even more so in the future. Think of carbon as negative money (debt): Too much of it is detrimental to the bottom line. In the Lean management system examined in Chapter 7, excessive carbon is appropriately considered a waste.

Managing the corporate carbon footprint requires a strategy that, when executed, can reduce the footprint. Formulating a strategy requires accurate data converted into useful information. In the case of carbon, an enterprise must inventory its carbon emissions and, in the language of the GHG Protocol, create an emissions profile. This profile must be comprehensive and accurate, because it is the basis of understanding business exposures and risks and, conversely, opportunities. It surfaces the carbon reduction opportunities, and facilitates the setting of reduction targets. The good news is that the know-how to achieving targeted reductions is growing rapidly, and shared in various forums.

The GHG Protocol manual lists other business goals served by a corporate GHG inventory:

- Participating in mandatory reporting programs. In many countries or in certain geographic regions, GHG reporting is required for corporate operations located in them. An accurate GHG corporate profile is required and assumed; it's the cost of doing business. Those organizations operating under a mandatory reporting situation have a potential competitive advantage over those that don't, in that the opportunities for reductions are known, and if acted on will reduce carbon, and in turn costs. An interesting possibility, with the maturing of the GHG Scope 3 (supply chain/value chain) measurement: Will the carbon footprint outside of the corporate geography be required in mandatory reporting?
- Public reporting and participation in voluntary GHG programs. Major institutional investors, controlling trillions of dollars, use a corporation's GHG reductions performance as part of their evaluation process before investing. This will be discussed later in this chapter. Another business goal served is favorable standing with the eco-sensitive buyer, from major corporations such as Wal-Mart, and the individual consumer. In some respects the voluntary reporting and active pursuit of reductions is a great marketing strategy.
- Participating in GHG markets. Without a carbon emissions inventory, a corporation couldn't buy needed carbon credits, or for those who are carbon frugal, sell them.

Hopefully the GHG markets are a temporary arrangement, and will over an extended period of time not be necessary as the cost of buying credits exceeds the cost of investment to reduce or eliminate them.

- Recognition for early voluntary action—early adopters are also thought leaders within an industry. Corporate customers, who also are early adopters, will give preference to suppliers who are leaders in carbon emission reductions. The expression "you are known by the company you keep" comes to mind. Another benefit for early adopters will occur as the GHG Scope 3 measurement is more widely implemented. Early adopters bring less carbon into the supply chain/value chain.

2. Setting Organizational Boundaries

Since the protocol differentiates between the direct and indirect generation of carbon emissions, the boundary of the organization must be established to make the determination between direct and indirect easier.

Because the protocol was developed to be compatible with corporate accounting and reporting principles, much of the thinking required to do this has already been done in many cases. Since many large agricultural conglomerates and food processors have diverse ownership and operational structures, criteria for determining the organization's boundary for ownership of Scope 1 emissions is required. The protocol has three criteria:

- Equity share
- Financial control
- Operational control

There are considerations that guide the application of the criteria:

- Economic substance takes precedence over legal ownership status.
- Who has the majority of risks and rewards of ownership of the operational assets?
- Which criteria best reflect the actual power of control?
- Who has full authority to introduce and implement operational policy?

Without an organizational boundary established, the important follow-on activities within the protocol can't be addressed, and also credible GHG accounting and reporting can't be done in the future.

The organizational boundary is very much like the foundation of a housing structure: It has to come first.

This section is meant to be a very brief overview, as there exists comprehensive guidance for establishing the organizational boundaries in the protocol guidance documents. These documents are readily available from the GHG Protocol website free of charge, and are written without jargon, and therefore easy to understand.

3. *Setting Operational Boundaries*

The next logical step in the protocol process is setting the operational boundaries. This is where agricultural and food processed products are manufactured, including all internal support services. Figure 8.15 shows the demarcation between the organizational and operational boundaries. At this boundary both direct (Scope 1) and indirect emissions (Scopes 2 and 3) are recognized.

One of the benefits of setting the organizational boundary is that it puts in focus Scope 3, the supply chains that are downstream of the organization. This is important in that companies often no longer compete as stand-alone entities, but on the basis of the supply chains they participate in.

Often the largest GHG emissions reduction opportunities reside upstream in a supply chain, and if a corporation has limited resources to bring to bear on its reduction efforts, and has reliable information on its upstream part-

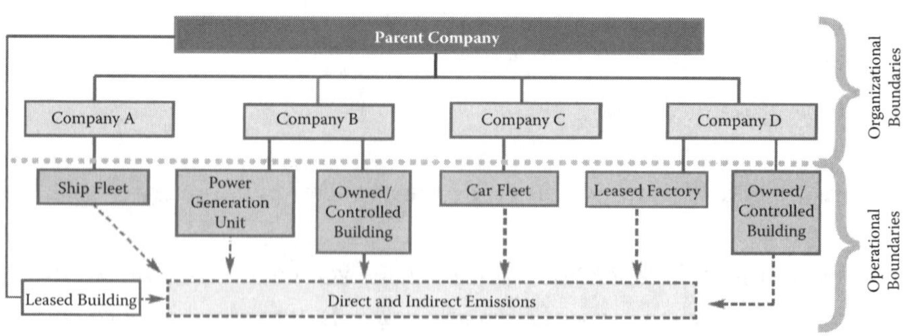

FIGURE 8.15
The demarcation between organizational and operational boundaries. (From *GHG Protocol Corporate Accounting and Reporting Standard*, revised edition. With permission.)

ners' GHG input, it can do a better job of allocating its limited resources by working with them to reduce their GHG product content.

For many organizations, the largest GHG culprit is purchased electricity (Scope 2) generated from nonrenewable raw materials such as coal. Unfortunately the power-generating infrastructure put in place many decades ago can't always be easily decommissioned and replaced with "clean" renewable sources. There are at least three viable strategies to deal with this reality:

1. The application of Lean techniques or continuous improvement techniques to reduce energy consumption. This was examined earlier in this chapter. The evidence to date is that it produces notable reductions and an excellent return on investment.
2. Redesign products to require less energy to process and produce.
3. Replace equipment that uses significantly less energy. Unfortunately this may require capital equipment investment and a long-term perspective.

A fourth possibility exists for large organizations, one that is a business innovation: buying energy from a green energy supplier. A noteworthy example is the Frito-Lay facility in Casa Grande, Arizona. This facility is something of a test bed for developing an eco-friendly factory. It contracted with a green energy supplier that built a dedicated power generating facility using solar technology. Frito-Lay made a long-term commitment, in turn getting electricity for a competitive rate, and avoided the large capital investment, and dramatically reduced its Scope 2, indirect carbon footprint. The supplier, which has the expertise to build and manage a solar energy generating complex, had a firm commitment with predictable cash flow, a win-win situation that also benefits a third party, citizens of this planet!

With the organization and operational boundaries defined, the framework to do GHG accounting and reporting is established. More work is needed, but the framework has been laid down (Figure 8.16). A

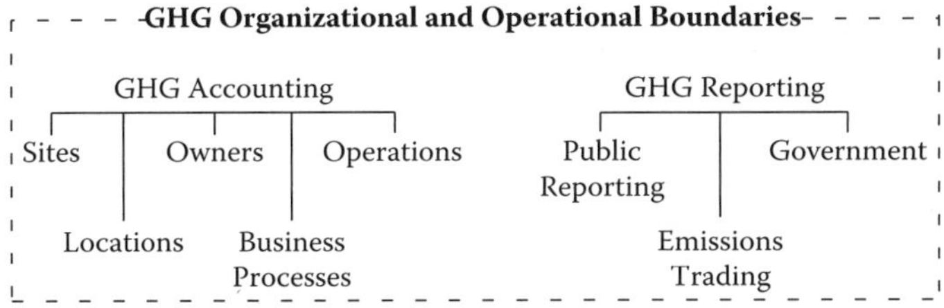

FIGURE 8.16
GHG organizational and operational boundaries.

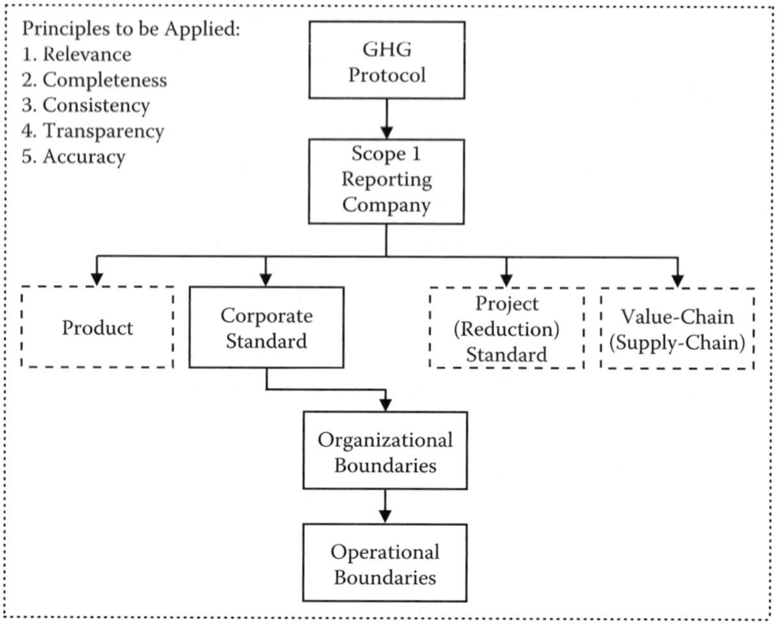

FIGURE 8.17
Summarizing the discussion of Scope 1 thus far.

mini-graphical overview of what has been covered to date is shown in Figure 8.17.

Identifying and Calculating GHG Emissions

The GHG Protocol methodology to identifying emissions within the defined operational boundary is straightforward for Scopes 1 and 2. The recommended process flow and sequence is:

- Identify sources.
- Select calculation approach.
- Choose emission factors.
- Collect data.
- Apply calculation tools.
- Roll up data to corporate level.

Identifying sources within Scope 1 requires thinking in terms of processes, products, and services. The typical emission sources are:

- Stationary combustion: Carbon generated as a by-product of combustion of fuel to heat boilers, furnaces, heaters, etc.

  Many manufactured food products require heat, and in some cases sustained heat. An example is an oil-based flavor that requires high cooking temperature, in a pressure vessel, over an extended period of hours, to create a reaction process.
- Mobile combustion: Carbon generated by fuels used to transport goods. An example is a gas-fueled forklift fleet to move raw materials and finished product within the factory.
- Process emissions: Emissions from physical or chemical processes that are often industry specific, and the GHG Protocol has supplemental guidance for these industries.
- Fugitive emissions: Unintentional (or intentional) releases from equipment leaks, or emissions from processes such as wastewater treatment. The distinction between what is a process versus intentional fugitive emissions has been established by experts, and codified by the GHG Protocol guidance.

GHG Scope 2 sources are even simpler to identify since it is purchased electricity. The emissions profile of the electricity provider is documented, as is how power is generated—coal, nuclear, natural gas, solar, or wind. Since many food processors are heavy consumers of electricity, the basis of how supply is generated is very important to its indirect emissions total, as is the effort to reduce electricity use within internal manufacturing processes.

Scope 3 is the sum of upstream and downstream supply chain/value chain contributors, and is the most recent addition to the protocol. The GHG Protocol has a well-defined methodology and is available to all contributors who generate direct emissions, and inherited indirect emissions. The challenge for the supply chain orchestrator or owner with Scope 3 isn't identification of probable sources, but collecting accurate information for reporting.

Select Calculation Method(s)

Once again the GHG Protocol does much of the figurative heavy lifting and provides guidance when it comes to selecting calculation methods. There is a hierarchy of catalogued calculating approaches that work, and can be examined for appropriateness to a corporation's situation. The most common approach is the application of documented emission factors that are calculated ratios, which function as a proxy measure of activity at an emission source.

Choose Emission Factors

Factors in this process step mean the basic raw source of emissions or a proxy estimate:

- Purchased fuels used directly within the operational boundary—Scope 1 (internally generated GHG emissions)
- Metered electricity used directly within the operational boundary—Scope 2 (purchased electricity, externally generated emissions)
- Published third-party emission factors—Scope 3 (externally generated emissions both upstream and downstream within the supply/value chain)

Collect Data

Because a structured process has been followed answering the what and how questions, it has systematically pointed to the data that must be collected, and with guidance from the GHG Protocol, the corporate accounting and reporting standard is made easy.

Roll-Up Data

Rolling up the GHG data is a bottom-up process very similar to a cost roll-up in a multilevel recipe/formulation or bill of material (BOM). The amount of specified GHG generated at the lowest-level source is passed on to the next higher level, and in turn that level summarizes all the inherited GHG from lower levels, plus the GHG it has generated, then summarizes both types (indirect and direct), and passes it to the next higher level. Conceptually it is identical to the system used to calculate water content in the Water Footprint Network methodology (virtual + direct water) (Figure 8.18). See Figure 8.19 for a mini-review.

Managing Inventory Quality

Managing inventory quality is the next step or consideration in the corporate accounting and reporting standard. Data quality is an excellent

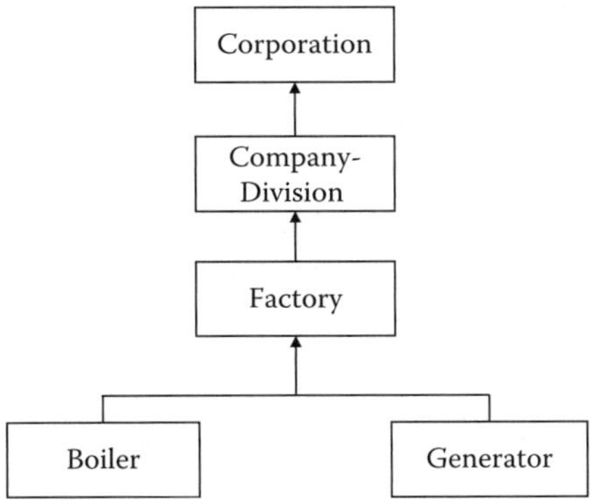

FIGURE 8.18
A simplified view of what is included in Scope 1.

indicator of organizational health. It is also critically important in food safety record keeping, and financial reporting for publically traded corporations. The protocol requires the establishment of a GHG inventory quality management system (Figure 8.20).

A collaborative partner in the GHG Protocol is the International Organization for Standardization (ISO), which has developed guidance on sustainability covering not only GHG, but also waste, water, and chemicals and toxins. The guidance starts with ISO 14001 and 14004, shown in Figure 8.21.

There is a specific set of ISO standards that directly support the GHG Protocol corporate accounting and reporting standard; they are 14064:1, 14064:2, and 14064:3. Figure 8.22 adds more detail to the previous one on the umbrella sustainability standard.

In practical terms the inventory quality management team should be established and chartered at the beginning of the GHG Protocol initiative.

Conference Room Pilot

Anyone who has ever introduced major change to processes understands that when data analysis and conceptual thinking meet hands-on usage head-on, there will be surprises! A conference room pilot, an approach used for decades to test assumptions, is the best way to avoid surprises and is highly recommended. The conference room pilot is also an opportunity to use the highly regarded quality methodology plan-do-check-act (PDCA), discussed in Chapter 7 on the Lean management system.

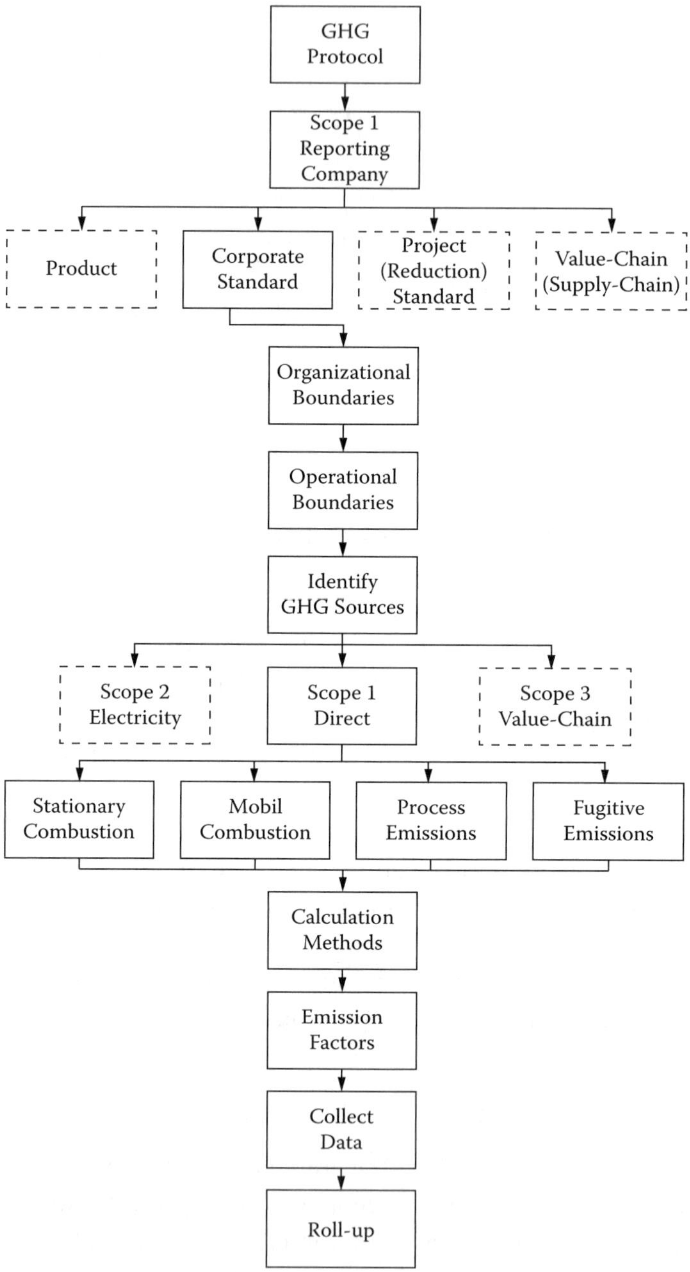

FIGURE 8.19
Progressively gaining an understanding of Scope 1 emissions.

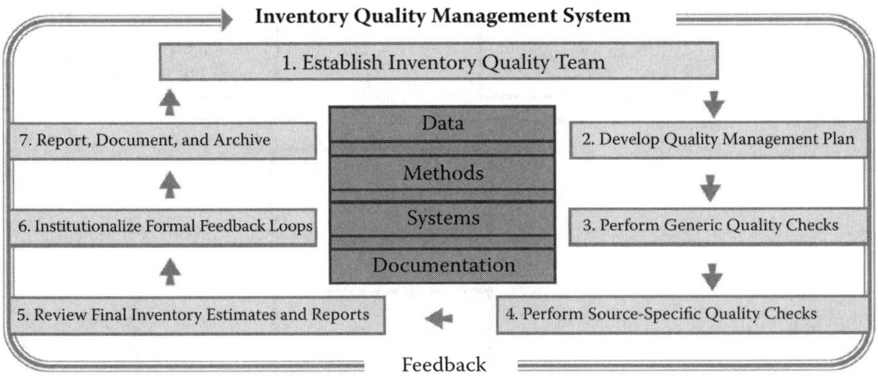

FIGURE 8.20
Inventory quality management system. (From *GHG Protocol Corporate Accounting and Reporting Standard*, revised edition. With permission.)

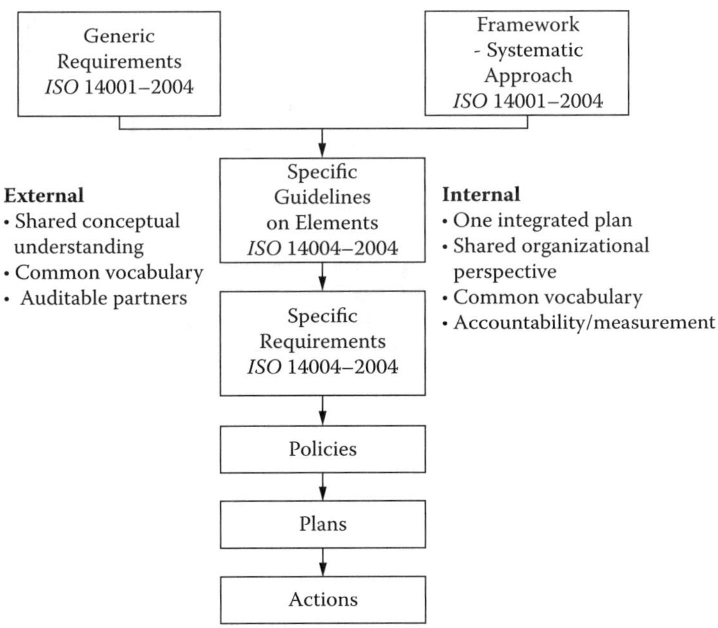

FIGURE 8.21
The ISO standards that support overall sustainability.

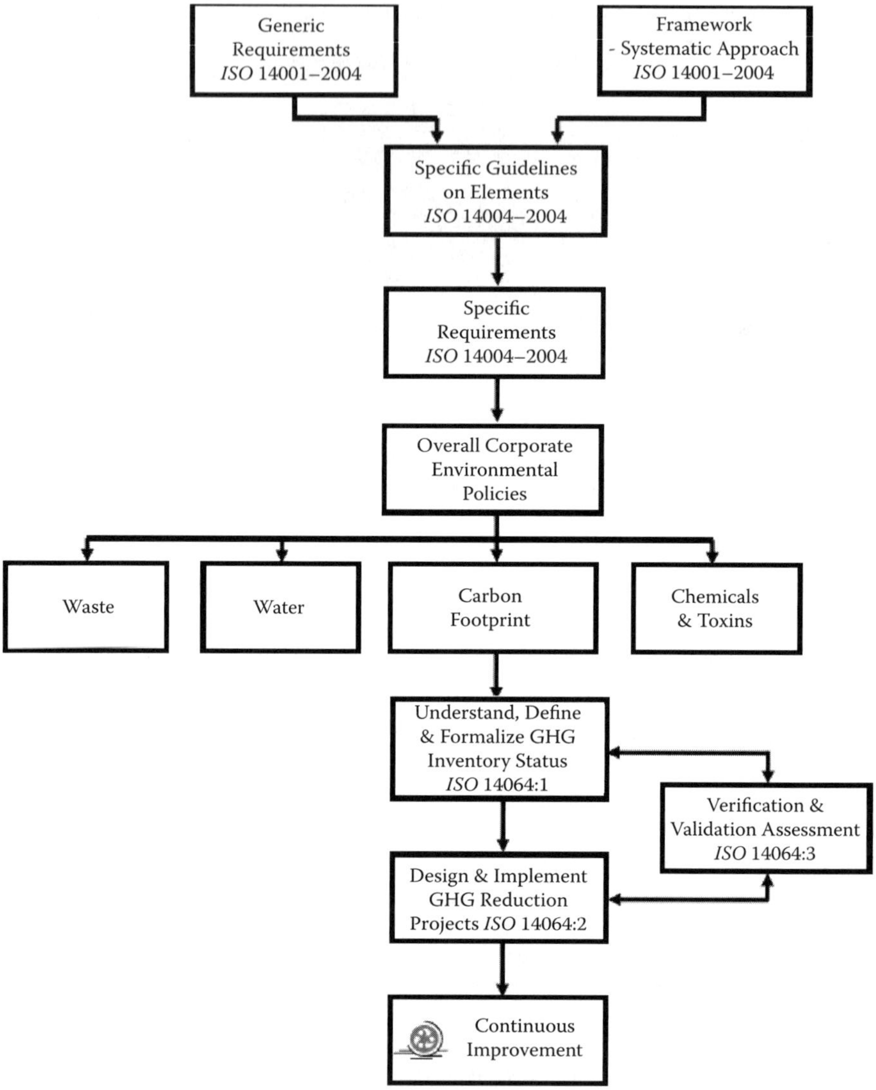

FIGURE 8.22
The full set of ISO standards related to GHG.

Now that:

- Strategic decisions have been made
- A quality system (ISO 14064) has been put in place
- Corporate and operational boundaries, rules, measures, etc., have been established

- Data have been collected and rolled up
- Assumptions have been tested through a conference room pilot

It's now time to set GHG reduction targets. People and organizations perform best with goals and targets established, and incentives put in place to achieve them. Setting reduction targets at first appears simple, but there is more to deciding the what and how much for organizations new to the GHG Protocol than one might think. Setting unrealistic or too-easy-to-achieve targets can be damaging to corporate reputation and organizational psychology. Fortunately the GHG Protocol provides guidance and forums to help make the target-setting process easier.

Setting a GHG Target

The first decision is which GHGs? The recommendation is one or more of the six GHGs covered by the Kyoto Protocol. The second is why: What are the drivers for setting the targets? There are at least seven possible drivers:

- Minimizing and managing GHG risks: Regulatory and reputation, potential loss of business revenue, and regulatory fines.
- Maximizing opportunities possible from GHG reductions: Generate new business revenue at the competition's expense.
- Cost savings: Energy generated from carbon sources is expensive, and will become even more expensive in the future.
- Stimulating innovation: The need to reduce GHG can (and should) drive innovation to develop replacement devices and services that are carbon neutral, thereby increasing revenue and market share.
- Preparing for future regulations: The pressures to decrease GHG emissions will continue to increase, and more stringent regulations are likely.
- Demonstrating corporate responsibility: The public will increasingly demand products and services from organizations with a viable GHG reduction history and current plans for further reductions.
- The Carbon Disclosure Project: A forum that addresses the needs of major institutional investors, who consider a corporation's environmental initiatives in their evaluation process.

As with anything that deals with implementing major change, the GHG Protocol guidance lists "ten commandments" in setting the GHG targets:

1. Obtain senior management commitment: Top-down commitment.
2. Decide on target type: This will be discussed below.
3. Decide on target boundary.
4. Choose the target base year: The year improvements will be measured against.
5. Define the length of the commitment period: Is this an annual or multiyear target?
6. Define the target completion date: If you don't have a destination, you'll never get there.
7. Decide on the use of offsets or credits: Offsets are a temporary fix at best. The ideal situation is to be carbon neutral and potentially have credits.
8. Establish a target double-counting policy: This is very important when using offsets (see number 7).
9. Decide on the target level: Constant small improvements or "big bang" large improvements or a mix of both?
10. Track and report progress: What doesn't get tracked and measured on a regular basis has a high risk of not achieving the target.

The first two considerations for setting the targets are the what and why; the third is deciding the target type. There are two target types, absolute and intensity:

- Absolute: A reduction over time in a specified quantity of a specific GHG.
- Intensity: A reduction in the relationship of a GHG emission relative to another business metric, i.e., sales, revenues, etc.

Both target types have advantages and disadvantages. An example of a disadvantage of the absolute type is when there is an increase or decrease in business revenue.

The last decade had a significant boom cycle (as opposed to a bust cycle). With increased business the enterprise might have significant increases in GHG emissions, followed by a decrease during the bust phase of the cycle. From a reporting standpoint it went from a normal (baseline) emitter, to an excessive emitter, followed by a misleading stellar GHG reduction achievement. Because of the strengths and weaknesses inherent in both types, some organizations use both in establishing their base year, targets, measurements, and reporting to more accurately reflect their GHG emissions reality.

There are other considerations: separate targets for operations that have different operating profiles, and if there are multiple geographically dispersed manufacturing or service units, can they provide accurate and reliable data? And finally, what scopes (1, 2, 3) should be considered for each operating unit. An example: A food flavors and seasonings manufacturer has two operating units:

- One makes liquid flavors that have complex formulations with many purchased ingredients. There is minimal capital equipment required, and no heating or cooling of the product. The manufacturing cycle is short. Scopes 1 (direct) and 2 (electricity) are easy; Scope 3 (supply chain) is the challenge.
- The other makes extracted liquid flavors produced under heat and pressure, with a simple formulation that often requires a multiday process, using expensive capital equipment with multiple equipment devices used. Since the products are heated, and both non-kosher and kosher products are produced, there is the need to use, based on the scheduling sequence of batches, both heat and water to kosherize the equipment. Scope 3 is easy (excluding machinery), and so are Scopes 1 and 2.

How constant (stable) the form of an organization is also influences what target type it selects for reporting in the event it needs to recalculate its performance based on acquiring or disposing of GHG consuming or generating business units. Companies with constant legal and operating assets probably would be candidates for the fixed choice. Those with change as part of their corporate persona would likely choose the intensity type because of the flexibility in recalculating performance. In the case of the enterprise that frequently reconfigures itself, using the intensity type for reporting makes sense, but using both fixed and intensity type concurrently might have merit.

Next up for consideration in establishing the GHG targets are the duration and completion dates. The Kyoto Protocol specifies a five-year first commitment period. Most companies use the single-year commitment approach. Both have merit; the five-year can accommodate unexpected events such as natural disasters. Anyone with management responsibility knows that managing the short-term numbers makes reaching the longer-term goals easier. The key consideration in the duration equation is the capital equipment (long-term) expenditure needed to reach goals.

Hopefully governments will treat these expenditures from an accounting perspective in a manner that will energize corporations to make these expenditures now versus later.

One of the ten commandments of setting a GHG reduction target is offsets. The GHG guidance states, "There are no generally accepted methodologies for quantifying GHG offsets." While the GHG Protocol covers offsets, it's alien to the intent of this book, which is to strive for excellence in food safety, operations, and supply-chain management. Offsets feel like financial derivatives. While OK selectively in the short term, they don't look promising in the long run unless there is a concrete plan to address the long-term problem in a meaningful manner. If the reader searches the Internet for attributes associated with business best practices, it is very unlikely "procrastination" will be found.

Much of the work on deciding the target level has been done in the previous steps, but there are other considerations:

1. Understanding the relationship between GHG emissions and business metrics
2. Relevant projected corporate growth factors that influence investment strategy and ROI
3. Existing corporate environmental initiatives that affect GHG emissions
4. Benchmarking GHG emissions with similar organizations
5. Future projected corporate GHG-related risks

Then develop multiple reduction strategies and simulate them to determine which produces the best outcomes. Following the simulation and identification of the best strategy, the next action item is the design of how to track and report progress internally in a consistent, complete, and transparent manner and which should include:

- Description of the target with a matrix of considerations: base year, completion date, etc.
- Information on emissions and performance in relationship to the target presented in a balance sheet-like format
- Carrying out regular performance checks

This step is an appropriate point to once again do a reality check with a conference room pilot. With all the relevant facts at hand, choose the target level.

Accounting for GHG Reductions

Now that the investigating, benchmarking, learning, verifying, forecasting, evaluating, selecting, and planning are done, and GHG reduction actions started, it's time to design the accounting structure for reductions. If this set of considerations suspiciously looks like a sub-set used in developing a business plan, it's because they are! There are two major realities that come into play that were considered in earlier phases:

1. The perspective of national and international policy makers
2. The bottom-up approach in the GHG Protocol that provides granularity in information in the following order: an individual source » facility » geographic cluster of operating units » corporate consolidation

The same level of care taken to establish a traditional general ledger (G/L) account structure for financial performance should be done for the GHG accounting, although possibly not the same degree of detail. Reporting for an organization that has relatively few changes in operating assets, such as acquisitions and divestures, will have less complexity in maintaining a baseline, and as a result in GHG accounting.

Reductions in Scope 2 (purchased electricity) and Scope 3 (supply chain/value chain), both indirect emission sources, do present some accounting challenges, because there often isn't a direct cause-effect relationship between the reporting organization and the resulting GHG emissions. But, the protocol guidance states that "concerns over accuracy should not inhibit companies from reporting their indirect omissions." With the recent enhancement in Scope 3 (supply chain/value chain), accuracy should improve over time. A rough estimate is better than no estimate.

There are examples of dramatic changes in Scope 2 consumption by organizations. One is a food company that collaborated with a solar electricity provider that built a generating complex, to be the sole provider of electrical energy to the company. The food company avoided capital expenditures to build the equivalent complex, and the Scope 2 provider had a secure long-term contract to supply. This type of GHG reduction would be covered under the protocol project quantification methodology (not covered in this book), using the demonstration of additionality to address this GHG accounting transaction. The type of reduction was from a project that resulted in emission reductions as well as those emissions that would have been generated had the project not been undertaken.

While there are other GHG reduction accounting-relevant factors, the last prominent one for food and supplement manufacturing is the avoidance of double accounting. The key principle is who controls the source of the GHG emission, and that ownership should be clarified to avoid double accounting.

Tracking GHG Emissions over Time and Accounting for Inventory Reductions

The last step before launching into a formal GHG Protocol commitment is choosing the base year against which progress will be measured and reported. This is not a trivial step.

Most companies select a single year, but a multiyear average may help smooth out unusual fluctuations, such as those a company might experience during the free market boom-bust economic cycles experienced in the past. Selection of the base year should relate to the business goals and sector in which the company operates.

Because change is inevitable, the protocol guidance covers the need to recalculate the base year so progress can be measured. While the need to recalculate is obvious for corporations that engage in activities such as mergers, acquisitions, and divestments, the same possibility exists for those engaging in significant out-sourcing or in-sourcing. In the case of out-sourcing and in-sourcing, the emissions move from Scope 1 to Scope 3 (out-sourcing), and from Scope 3 to Scope 1 (in-sourcing). The protocol guidance covers other changes so the "how to" to recalculate the base year is codified, and knowledgeable resources are readily available to assist if needed.

Verification and Reporting of GHG Emissions

What will be reported was defined earlier in the GHG emission management process, tracked, and recorded in a disciplined manner as outlined in the inventory quality plan.

Reporting isn't difficult, but it would be prudent to have the report examined not only by senior management, but also by the corporate board of directors (BOD), and then through an external verification before publishing it in the GHG Protocol, and the outside world. When the Carbon Disclosure Project is discussed, the need for accuracy, clarity, and transparency will become obvious.

The required information is similar to that required by many investors, or perhaps even important customers, who want to examine an organization's business plan and progress against goals. There is nothing esoteric in the required and voluntary supplemental information.

The basics: Scope 1 and 2 reporting constitutes the minimum to be reported, Scope 3 is optional, and the GHGs from the Kyoto Protocol that are excluded from the report are listed. Figure 8.23 highlights the information that should be in the report.

Verification prior to reporting of GHG status is the norm, and in many respects is similar to the independent audit by a qualified independent organization of the financial records of a company, in support of their annual financial statements. The objective is to determine the overall inventory data quality, and the systems influencing it. Two contributors to the overall data quality are the company's inventory quality systems and adherence to ISO 14064, both of which were discussed earlier in this chapter.

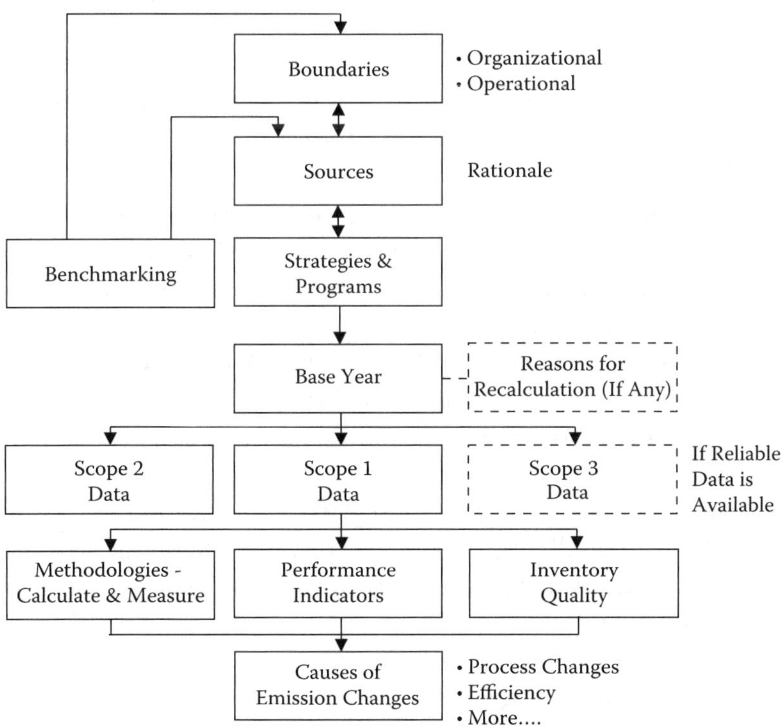

FIGURE 8.23
The key elements in a GHG Protocol report.

At the beginning of a corporation's GHG emissions reduction commitment, the goals associated with it were established, refined and validated as the effort progressed, and should be revisited for further evaluation before the independent verification.

This is very much in keeping with the plan-do-check-act (PDCA) quality management thinking that supports continuous improvement. Common goals cited in the GHG guidance document are the increasing of:

- Credibility with stakeholders on the organization's sustainability initiatives
- Confidence in GHG reduction target setting
- Senior management confidence in major investment decisions that answer the question: Do they support continuous improvement in corporate sustainability?
- Learning, knowledge, and skill in internal accounting and reporting practices
- Preparation for current or future mandatory verification requirements

After this reexamination of goals, the scope of the independent verification should be determined, and what will be included and excluded and how the verification will be conducted, and a clearly defined scope of work (the flipside of a statement of work) developed.

As with financial auditing, the concept of materiality comes into play during the verification process. Information is considered material (significant) if by its exclusion or inclusion it influences the possible actions by the reader-analyzer. Since the concept of materiality involves a value judgment such as how big a deviation should be before it crosses the boundary of accuracy (materiality threshold), it is often established by specific GHG programs or national requirements and is expressed as a percentage, e.g., five percent.

Before the audit, the verifier needs to plan the verification process based on potential material discrepancy risk, which is based on the soon-to-be-audited organization's industry sector, organizational structure, and complexity of operations.

Also, the usual consideration for anything related to organizational change, determining management commitment, and involvement is incorporated into the plan. Much of the audit will be focused on the inventory quality management system's completeness and integrity. Figure 8.24 is the simple process flow of pre- and post-verification.

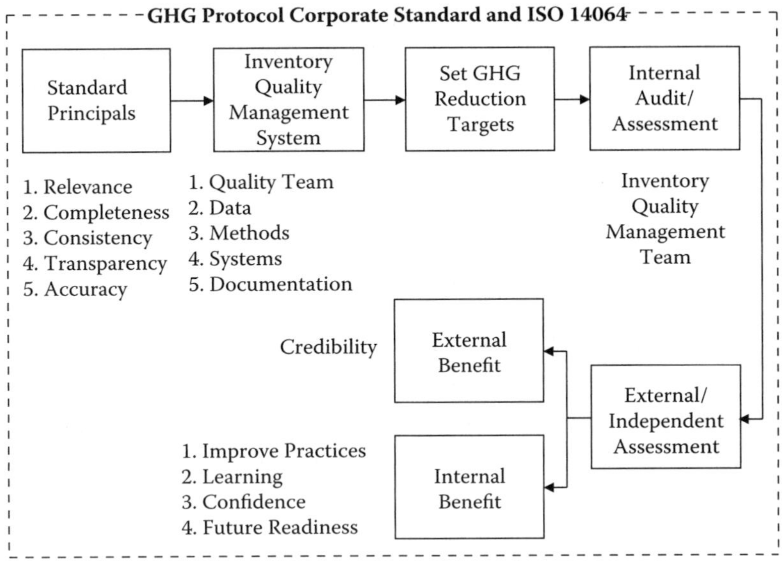

FIGURE 8.24
GHG Protocol corporate standard and ISO 14064.

Before leaving the discussion of the GHG Protocol, it should be noted that three other guidance documents haven't been discussed as they are beyond the scope of this book, but are readily available for examination and study:

1. Project (quantification of reductions)
2. Product life cycle
3. Value chain (supply chain)

Carbon Disclosure Project (CDP)

The Carbon Disclosure Project (CDP) website states its mission as follows:

> The Carbon Disclosure Project (CDP) works to transform the way the world does business to prevent dangerous climate change and protect our natural resources. We see a world where capital is efficiently allocated to create long-term prosperity rather than short-term gain at the expense of our environment.
>
> Evidence and insight is vital to driving real change. We use the power of measurement and information disclosure to improve the management of environmental risk. By leveraging market forces including shareholders,

> customers and governments, CDP has incentivized thousands of companies and cities across the world's largest economies to measure and disclose their greenhouse gas emissions, climate change risk and water strategies. We put this information at the heart of business, investment and policy decision making.

A few key thoughts from this mission statement:

- Transform the way the world does business.
- Protect our natural resources.
- Capital is efficiently allocated to create long-term prosperity.
- Use the power of measurement and information disclosure.
- Put this information at the heart of business, investment, and policy decision making.

CDP focuses on both GHG emissions and water, and utilizes methodologies and standards from other credible organizations, such as the GHG Protocol. Its underlying logic is that investment money follows opportunity and avoids risk. Lack of sustainability or the maintaining of the current status quo brings with it substantial risk, and constrains innovation and opportunity. The key is making available to institutional investors usable information on corporate sustainability plans and progress, using accepted methodologies to quantify and measure. Many investors and financial analysts regard good performance on environmental and social measures as a proxy for good management.

What CDP has developed for its members is the world's largest database on a corporation's sustainability plans and the progress against them that is standardized, and has provided web-based systems to access this database.

The database is part of an information platform that also includes an input questionnaire, benchmarking tools, analytics, etc. Because of its completeness, institutional investors representing over U.S.$78 trillion utilize this service and use CDP as a facilitator to request sustainability information from corporations. The corporate world has responded to this opportunity, and over eighty percent of Global 500 corporations have submitted reports to the CDP database. Most of the reporting to date has been on Scope 1 and 2 emissions.

Even before the release of a finalized Scope 3 guidance, a supply chain initiative was launched by members of the CDP.

Members from the food industry include:

- Coca-Cola
- Con Agra Foods
- Kraft Foods
- Nestle
- PepsiCo
- Unilever
- Major supermarket chains such as Wal-Mart

It is forecasted that numbers of large food manufacturers and brand owners will increase the number of participants in the supply chain initiative.

A coauthored report from the CDP Supply Chain Member Program (CDP Supply Chain Report 2012) and Accenture (a respected consulting and services company) presented a current status of Scope 3 progress and trends based on industry knowledge and a survey. One of the information points in the report centered on the climate: "Climate scientists predict that storms will become more frequent and more intense in the future leading to potentially greater risks for companies with extended supply-chains or agricultural networks (or both)" (p. 5). The respondents seemed to have read the climate scientists' assessment because:

- More than half (fifty-three percent) of respondents expect climate change to lead to increased operational costs, and seventy-four percent of those that cited regulation as a risk expect increased costs to impact their business within the next five years.
- Ninety-four percent of them have incorporated climate change into their company-wide risk management process.
- CDP supply chain member companies who are requesting climate information from their suppliers—ninety percent of responding companies—have a climate change strategy with at least general guidelines for procurement.

A trickle-down or push-down effect has started. The significant percentage of submissions by large corporations to the CDP database that have developed better environmental practices in response to demand from consumers and investors now have started to focus on suppliers to further reduce GHG emissions. In response suppliers are now beginning to improve their own practices in response to demand from corporate clients.

This is an appropriate response because corporations that have succeeded in managing their own internal operations are now shifting their business models to drive long-term change along the entire value chain to:

1. Increase revenues
2. Reduce costs (carbon and water are expensive)
3. Maintain or improve brand reputation
4. Improve risk management

This shift is evident: The number of companies that actively train their procurement staff in supply chain carbon management is increasing steadily:

- 2011—63%
- 2010—42%
- 2009—26%

With that rate of growth, significant change is on the horizon—change such as the inclusion of a contractual obligation for suppliers to manage GHG emissions, and include information about emissions management in Request for Proposal (RFP) responses. Another indicator is that a significant number of survey respondents indicated they will soon begin deselecting suppliers that do not adopt such measures.

One issue mentioned in the CDC report is the monetization of GHG reduction efforts by the supplier community. Fortunately this will become easier in the future with the publication of the GHG Protocol Scope 3 standard, which clearly defines fifteen categories of Scope 3 emissions and provides detailed guidance on how to account for and report those emissions. An example used in the report follows:

> Kraft Foods, a global food products company, developed a complete scope 3 inventory to understand and evaluate its value chain GHG emissions. In the first year of scope 3 reporting, Kraft Foods used industry-average life cycle inventory data from various public and commercial sources to calculate scope 3 emissions.
>
> The company found that scope 3 emissions comprise more than 90% of the company's combined scope 1, scope 2, and scope 3 emissions. Within scope 3, Kraft Foods found that emissions from category 1 (purchased goods and services), including raw materials and its agricultural supply-chain, comprised 70% of its total scope 3 emissions. Kraft Foods plans to continuously improve the quality of its GHG inventory.

> The company's experience highlights the value of using secondary data to identify where to prioritize GHG reduction efforts and target opportunities to collaborate with suppliers in future years to better measure progress and achieve GHG reductions.
>
> **—Pankaj Bhatia, Director, GHG Protocol, World Resources Institute, p. 6**

One disturbing point for North American-based companies noted in the report is that suppliers in Asia and Europe are more likely to have comprehensive climate change strategies in place—and have generated better results—than their competitors in North America. The disclosure and performance scores of Asian-based and European-based companies exceed the scores of companies from North America. This fact, in combination with the views of the U.S. FDA, should be a wake-up call:

> In the decade ahead, the world economy will be shaped by several distinct forces: the rise of emerging markets, **the scarcity of natural resources**, and the increased flow of capital, information and goods across borders.
>
> **—*Pathway to Global Product Safety and Quality* (a special report), U.S. Food and Drug Administration (FDA)**

From a supplier management standpoint, the next steps are to more effectively evaluate suppliers, improve performance through more effective procurement, and improve the tools and metrics used to quantify and monetize the gains from emissions reductions. With the maturing of the GHG Protocol Scope 3 this will become progressively easier.

In summary, the CDP and its institutional members have exerted significant influence with the Global 500 corporations, which in turn will put emphasis on their supply chain member GHG emission reduction efforts, because they do produce business benefits.

Mini-Summary Carbon

- Emissions from carbon sources are expensive and will most likely increase in cost in the future, and represent significant corporate risk.
- How to understand, inventory, set up accounting for, and reduce GHG emissions is codified, and the how to readily available through the GHG Protocol, which includes Scope 1 (direct),

Scope 2 (purchased electricity), and Scope 3 (supply chain/value chain) segmentation.
- The GHG Protocol methodologies have been accepted by the majority of Global 500 corporations, and have become the *de facto* standard for GHG accounting and reporting.
- The accounting and reporting on GHG has many similarities with traditional financial reporting for publicly traded companies.
- ISO has collaborated with the GHG Protocol to create specific GHG accounting standards.
- Most GHG reduction emphases to date have been on Scope 1 (direct) and Scope 2 (purchased electricity).
- The finalization of Scope 3 and the critical mass of early adopters of Scopes 1 and 2 will put more emphasis on Scope 3.
- The CDP has the largest database of GHG reports in the world, and over eighty percent of the Global 500 participate and report GHG status. It also has the support of large institutional investors who use the reporting in their evaluation of potential investments. CDP utilizes the GHG Protocol methodology.
- Within CDP there is a supply chain group of members focusing on the supply chain in a business model shift, who are now looking to suppliers within their supply chains to report and put together a GHG emissions reduction plan. This effort is gaining momentum, and procurement practices are undergoing change. Facilitating this is the availability of a workable Scope 3 guidance.
- There is a need for a wake-up call within the North American supplier community, as both Asian and European suppliers seem more prepared to manage GHG emissions and participate in the process.
- Carbon and water both have a significant impact on food manufacturing.
- This section on carbon is at best an orientation to GHG management, and the reader is encouraged to do further study.

THE UN GLOBAL COMPACT AND FAIR TRADE SUPPLY

The UN Global Compact

The contribution to the sustainability movement within the business community by the United Nations is often not appreciated. It has progressively expanded the awareness within the business community of the impact of GHG emissions on the planet. With awareness and subsequent investigation, corporations discovered the high cost of GHG emissions, and now are becoming aware of water availability as a business risk.

It all started in 1972 during a difficult period of time in which the world was divided into three political and economic groupings: capitalist/free market, communist/socialist, and nonaligned or neutral countries. Very difficult times to foster cooperation and instigate change, and it took almost twenty years (Rio Declaration) to get beyond the awareness and vision stage to defining actions necessary to work toward sustainability. Following that it only took five years for countries on a country-by-country basis to commit to GHG reductions (Kyoto), with some abstentions. At this point two factors came into play:

- The end of the Cold War, with the rapid movement toward interconnected global free trading
- The momentum of the UN sustainability initiatives

Both in combination led to the creation of the UN Global Compact directed toward corporations that now wanted to trade and invest globally. Since its inception in the year 2000, the compact has had rapid corporate adoption.

Currently it is the largest voluntary corporate responsibility initiative in the world. The Global Compact provides collaborative solutions to challenges facing both business and society by combining the best properties of the UN with the private sector's solution, finding strengths utilizing the compact's ten principles in business activities, and applying its expertise. The UN website states: "The Global Compact is global and local, private and public, voluntary yet accountable," and also that the "Global Compact is a complement to regulatory regimes," not a replacement for them.

The compact is focused on four responsibilities:

- Environment
- Human rights
- Labor
- Anticorruption

The environmental component is the by-product of the thirty years of UN experience in working with sustainability issues and what is needed to mitigate them. Human rights speaks to the morals of human treatment, and anticorruption to the ethics of an interconnected global trading community that has the need to level the playing field in competing for business.

Within these four groupings there are ten principles, of which three cover sustainability:

- Principle 7: Businesses should support a precautionary approach to environmental challenges.
- Principle 8: Undertake initiatives to promote greater environmental responsibility.
- Principle 9: Encourage the development and diffusion of environmentally friendly technologies.

Principle 7 has its origins in the 1992 Rio Declaration (Agenda 21), which takes global warming doubters head on:

1. Where there are threats of serious or irreversible damage, lack of full scientific certainty shall not be used as a reason for postponing cost-effective measures to prevent environmental degradation
2. Precaution is applied as long as scientific information is incomplete or inconclusive and the associated risk is still considered too high
3. The idea of prevention rather than cure—it is more cost-effective to take early action … long-term returns (for) investing in sustainable operation (means) less financial risk
4. Join industry-wide collaborative efforts to share knowledge and deal with issues, in particular production processes and products around which high level of uncertainty, potential harms and sensitivity exist

The next principle, number 8, also has its foundation in the Rio Declaration (Agenda 21) and explicitly encourages self-regulation:

> Business and industry should increase self-regulation, guided by appropriate codes, charters and initiatives integrated into all elements of business planning and decision making, and fostering openness and dialogue with employees and the public.

The how is also defined in the Rio Declaration:

- Develop sustainability targets and indicators (economic and environmental).

- Establish a sustainable production and consumption program with clear performance objectives to take the organization beyond compliance in the long term.
- Work with suppliers to improve environmental performance, extending responsibility up the product chain and down the supply chain.
- Adopt voluntary charters, codes of conduct, or practice internally as well as public initiatives to confirm acceptable behavior and performance.
- Measure, track, and communicate progress in incorporating sustainability principles into business practices, including reporting.
- Ensure transparency and unbiased dialogue with stakeholders.

In this guidance, emphasis is placed on the existence of appropriate management systems being crucial. Figure 8.25 highlights the influence of the UN Global Compact and how thought leaders within industry and other stakeholders have responded to it.

The final principle that deals with sustainability is principle 8: Encourage the development and diffusion of environmentally friendly technologies.

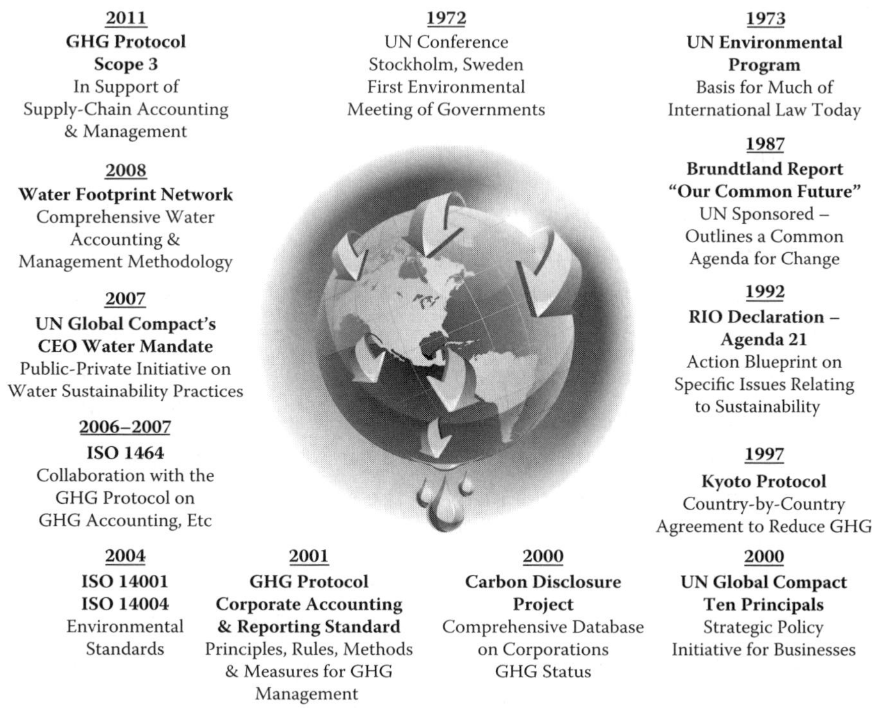

FIGURE 8.25
The subtle but powerful influence of the UN Global Compact.

This is both a risk and a reward situation. If a corporation's products and services aren't moving in the direction of being GHG neutral, the risk of failure is high because a competitor can gain market share through the customers' preferences, as well as producing at a lower cost. The analogy and assumed truths of the turtle versus hare race don't apply here; the GHG emissions reduction turtle loses.

There is a flip side to the risk, and that is the reward. Significant parts of our hard infrastructure, both civil and consumer based, will be replaced because of high operating costs and GHG regulations. Those early to market with eco-friendly products and services will have advantages.

An organization committing to the UN Global Compact assumes the responsibility to report on all four groupings. This annual report is called Communication on Progress (COP), and each and every report submitted is reviewed. All reports are available to the public. If an organization is a participant in the CDP the reporting on sustainability will be easy, but there are other reporting areas (labor, etc.) that will require due diligence. A COP has a life span of one year, and an up-to-date COP must be submitted annually.

Fair Trade

Fair trade is generally associated with developing countries that provide agricultural products as exports to developed countries. There is a sustainability aspect to fair trade; an example is the Aral Sea disaster. Although it's not a food example, it is similar to many food production situations in that water was consumed for an export product that was ecologically unsupportable. In this case the consequences are more than just economic; thousands of young children are dying from the particulate pollution that resulted from the draining of the sea. Figure 8.26 is better than a thousand words.

What fair trade also brings into play is the social justice aspects of the UN Global Compact with regards to child labor, labor relations, and possibly forms of slavery. While GHG emissions have most of the corporate world's attention, and water issues are gaining momentum, fair trade will over time become an increasing concern for food industry supply chain executives.

FIGURE 8.26
Cotton for export—former Aral Sea, Central Asia. (From Arjen Hoekstra, Water Footprint Network. With permission.)

SUMMARY

As the U.S. FDA special report referenced earlier emphasized, the future holds issues regarding sources of supply for the food and nutritional supplements industry. Also, the food supply chain has become global with many new competitors.

Sustainability, for both water and GHG-emitting energy, has become expensive, and is forecast to become more so. The twin negatives of undersatisfied human needs for water and energy and the increasing costs to make food products because of those needs require an additional focus on waste reduction. Adoption of the Lean system of thinking, which is centered on the elimination of waste, needs to be applied to this challenge.

The availability of fresh usable water is now finally being recognized as a current or future problem. Fortunately the United Nations has made this a priority, and new insightful water management accounting strategies and a comprehensive water management methodology have been developed, most notably the Water Footprint Network methodology. Water is at the very heart of the food supply chain, starting with agriculture, and in the future will no longer be cheap.

GHG emission sources have been recognized as a two-headed problem: climate change with the disruptions it is forecasted to bring to both the global society and the economy, and the seemingly ever-increasing cost of carbon-based/emitting fuels. In retrospect, the United Nations initiatives were prescient, and what was started in the early 1970s by the UN was the catalyst for the development of the following:

- UN Global Compact
- GHG Protocol (Scopes 1 and 2)
- ISO sustainability standards
- Carbon Disclosure Project
- UN Global CEO Water Mandate
- Water Footprint Network
- GHG Protocol (Scope 3: supply chain/value chain)

Both food and nutritional supplements manufacturers are water and energy intensive, if not directly, most likely due to water and GHG costs inherited through their supply chains. The second decade of this millennium will be the sustainability decade.

QUIZ

8.1. The current driver on sustainability is solely for ecological (green) considerations.
☐ True ☐ False

8.2. The U.S. FDA special report *Pathway to Global Product Safety and Quality* forecasts adequate supply of natural raw materials.
☐ True ☐ False

8.3. Recycling is the only method to reduce physical landfill waste.
☐ True ☐ False

8.4. Products made from recycled materials always use no water or energy resources.
☐ True ☐ False

8.5. Early adopters of sustainability measures will be focusing on their supply chains in the near future.
☐ True ☐ False

8.6. There are no current tools/methodologies to model and manage an end-to-end supply chain.
☐ True ☐ False

8.7. Water is rapidly becoming a scarce resource within certain geographic regions.
☐ True ☐ False

8.8. What is the Water Footprint Network methodology?

8.9. What are the three types of water within the Water Footprint Network methodology?

8.10. During the manufacture of a food product what other two types of water are involved?

8.11. What global organization has focused attention on water management?

8.12. Reducing energy consumption can only be accomplished through the acquisition of next-generation capital equipment to replace energy-intensive equipment.
☐ True ☐ False

8.13. What was the UN landmark sustainability initiative that involved the private sector?

8.14. What two major organizations have responded to the UN initiative for sustainability for GHG emissions management?

8.15. What organization has developed the *de facto* guidance on GHG accounting and reporting?

8.16. Are there standards that support the guidance on GHG accounting and reporting?
☐ Yes ☐ No

8.17. What are GHG accounting and reporting conceptually similar to those of a publicly traded corporation required to do?

8.18. What are the three distinct types of emissions generation categories that are contained within the GHG accounting and reporting methodology?

8.19. Which one of the categories is currently considered optional?

8.20. Is there a requirement for absolute precision in reporting for corporations new to the GHG accounting and reporting processes?
☐ Yes ☐ No

8.21. What is the Carbon Disclosure Project (CDP)?

8.22. Name two major stakeholders who utilize the CDP's capabilities.

8.23. What is the name of the UN initiative on water management and who is it addressed to?

8.24. With the finalization of GHG Protocol's Scope 3, what will early adopter large food retailers and large food companies concentrate on?

The answers to these questions can be found in the Appendix along with definitions, website addresses, and recommended reading.

Appendix

The appendix is organized as follows:

1. Glossary
2. Quiz answers
3. Food safety
4. Magazines focused on operations in the food industry
5. Emerging food-related technologies
6. Professional associations that support excellence in operations and supply-chain management
7. Professional practitioner certifications
8. Sustainability
9. Recommended reading
10. Miscellaneous websites
11. Reference List

Please note that while it's unusual, the website addresses referenced can change.

GLOSSARY

The following definitions come from a variety of sources. These include the *APICS Dictionary*, a dictionary of Lean terms compiled by Mike Osterling (© Karen Martin and Mike Osterling, 2010. With permission where applicable.), and other credible sources. Definition sources are noted at the end of the definition including Water Footprint Network, Free Medical Dictionary, Free Dictionary, FDA (Federal Drug Administration), About.com, EFSA (European Food Safety Authority), Mayo Clinic, GHG Protocol, IFT (Institute of Food Technologists), ASQ (American Society for Quality), NSF, The State of California, FSIS USDA (Food Safety Inspection Service, United States Department of Agriculture), WHO (World Health Organization), and CIO (Chief Information Officer) Magazine.

6S: An extension adding safety to the Lean method 5S. 5S is intended to provide workplace organization and housekeeping. The term *5S* derives from a list of five Japanese words: *seiri*, *seiton*, *seiso*, *seiketsu*, and *shitsuke*. Translated, these words mean proper arrangement (*seiri*), orderliness (*seiton*), cleanliness (*seiso*), cleaned up (*seiketsu*), and discipline (*shitsuke*). The 5Ss are performed in this order: sort, straighten, sweep, schedule, and sustain.

A3: A problem-solving methodology (and format) developed by Toyota that provides a methodology for the PDCA process. Osterling.

AI (artificial intelligence): (1) Computer programs that can learn and reason in a manner similar to humans. The problem is defined in terms of states and operators to generate a search space that is examined for the best solution. In contrast, conventional programming collects and processes data by algorithm or fixed step-by-step procedures. (2) An area in computer science that attempts to develop AI computer programs. APICS.

Ambient water quality standards: The maximum allowable amount of a substance in rivers, lakes, or groundwater, given as a concentration. Ambient water quality standards can also refer to other properties of the water, such as temperature or pH. Standards are set to protect against anticipated adverse effects on human health or welfare, wildlife, or the functioning of ecosystems. Water Footprint Network.

AME (Association for Manufacturing Excellence): A nonprofit professional association dedicated to the exchange of knowledge in organizational excellence through the implementation of techniques such as Lean tools, leadership, and Lean product development. Lean supply chain and Lean accounting.

APICS (Association for Operations and Supply Chain Management): Founded in 1957 as the American Production and Inventory Control Society, APICS builds operations management excellence in individuals and enterprises through superior education and training, internationally recognized certifications, comprehensive resources, and a worldwide network of accomplished industry professionals. APICS.

***APICS Dictionary*:** The *de facto* lexicon of supply-chain and operations management that has more than 4,000 standard terms and definitions.

APICS OMBOK (Operations Management Body of Knowledge) Framework: A respected publication that defines the scope of operations and supply-chain management profession, and is a resource to understand operations and supply-chain management today and in the future. The APICS OMBOK is updated annually.

APS (advanced planning system): Techniques that deal with analysis and planning of logistics and manufacturing during short-, intermediate-, and long-term time periods. APS describes any computer program that uses advanced mathematical algorithms or logic to perform optimization or simulation on finite capacity scheduling, sourcing, capital planning, resource planning, forecasting, demand management, and others. These techniques simultaneously consider a range of constraints and business rules to provide real-time planning and scheduling, decision support, and available-to-promise and capable-to-promise capabilities. APS often generates and evaluates multiple scenarios. Management then selects one scenario to use as the official plan. The five main components of APS systems are (1) demand planning, (2) production planning, (3) production scheduling, (4) distribution planning, and (5) transportation planning.

ASQ (American Society for Quality): A not-for-profit educational organization with more than 100,000 individual and organizational members who are interested in quality improvement. APICS.

ASTD (American Society for Training and Development): The world's largest association dedicated to the art and science of workplace learning and development. In recent years ASTD has widened the organization's focus to link learning and performance to individual and organizational results.

Bar coding: A method of encoding data using bar codes for fast and accurate readability. APICS.

Batch process ERP (Enterprise Resource Planning): An enhanced version of ERP designed to support organizations making formula- or recipe-based products such as food, nutritional supplements, pharmaceuticals, and specialty chemicals. This enhancement includes subsystems supporting regulatory safety requirements, including lot tracing, HACCP, allergen management, SARA reporting, and MSDS.

Benchmarking: Comparing a company's costs, products, and services to those of a company thought to have superior performance. The benchmark target is often a competitor but is not always a firm in the same industry. Seven types of benchmarking have been cited: (1) competitive benchmarking, (2) financial benchmarking, (3) functional benchmarking, (4) performance benchmarking, (5) process benchmarking, (6) product benchmarking, and (7) strategic benchmarking. APICS.

Best practices: In benchmarking, the measurement or performance standard by which similar items are evaluated. Defining a best practice identifies opportunities to improve effectiveness. The process of comparing an actual result to a best practice may be applied to resources, activities, or cost objects. APICS.

Bioterrorism Act: A law passed by the U.S. Congress to enhance the security of the United States. The act is divided into five titles, with a title directed at the food industry: "Protecting Safety and Security of Food and Drug Supply." The implementation of the law was the responsibility of various government agencies such as the FDA.

Biotoxin: A toxic substance produced by a living organism. *Free Medical Dictionary.*

Blue water: Fresh surface and groundwater, i.e., the water in freshwater lakes, rivers, and aquifers. Water Footprint Network.

Blue water footprint: Volume of surface and groundwater consumed as a result of the production of a good or service. Consumption refers to the volume of freshwater used and then evaporated or incorporated into a product. It also includes water abstracted from surface or groundwater in a catchment and returned to another catchment or the sea. It is the amount of water abstracted from ground- or surface water that does not return to the catchment from which it was withdrawn. Water Footprint Network.

BOM (bill of material): (1) A listing of all the subassemblies, intermediates, parts, and raw materials that go into a parent assembly showing the quantity of each required to make an assembly. It is used in conjunction with the master production schedule to determine the items for which purchase requisitions and production orders must be released. A variety of display formats exist for bills of material, including the single-level bill of material, indented bill of material, modular (planning) bill of material, transient bill of material, matrix bill of material, and costed bill of material. (2) A list of all the materials needed to make one production run of a product, by a contract manufacturer, of piece parts/components for its customers. The bill of material may also be called the formula, recipe, or ingredients list in certain process industries. APICS.

Carbon Disclosure Project (CDP): An independent not-for-profit organization focused on helping companies and cities achieve greenhouse gas emissions reduction and sustainable water practices. CDP maintains an electronic database of annual reports from thought-leading corporations and cities conforming to the GHG Protocol. It also functions as a request broker between large institutional investors and corporations, and large corporations such as Wal-Mart and their supplier community.

Carbon tax: A tax on the emissions caused by the burning of coal, gas, and oil, aimed at reducing the production of greenhouse gases.

Carrying cost (of inventory): The cost of holding inventory, usually defined as a percentage of the dollar value of inventory per unit of time (generally one year). Carrying cost depends mainly on the cost of capital invested as well as such costs of maintaining the inventory as taxes and insurance, obsolescence, spoilage, and space occupied. Such costs vary from ten to thirty-five percent annually, depending on type of industry. Carrying cost is ultimately a policy variable reflecting the opportunity cost of alternative uses for funds invested in inventory. *Synonym:* Holding costs. APICS.

CARVER + Shock: An offensive targeting prioritization tool that has been adapted for use in the food sector. This tool can be used to access the vulnerabilities within a system or infrastructure. CARVER is an acronym for Criticality, Accessability, Recuperability, Vulnerability, Effect, Recognizability. The modified CARVER tool evaluates a seventh attribute, the combined health, economic, and psychological impacts of an attack, or the shock attributes of an attack. FDA.

CCP (critical control point): An integral component of an HACCP analysis and plan. A step at which control can be applied and is essential to prevent or eliminate a food safety hazard or reduce it to an acceptable level. FDA.

CDC (Centers for Disease Control): The CDC's main functions include monitoring public health, detecting and investigating health problems, conducting research to prevent health problems, developing and advocating public health policies, implementing prevention strategies and measures, promoting healthy lifestyles and behavior, fostering safe and healthy environments, and providing leadership, education, and training to enhance public health.

Cell (layout for flow): The co-location of processes or equipment in sequence to permit one-piece flow and the flexible deployment of workers to operate multiple processes (resources). The resources found in cells are often cross-functional in nature. Also referred to as cells or cellular arrangement. Contrast with functional arrangement. Osterling.

Certificate of analysis (C of A): *See* C of A.

cGMP (current good manufacturing practices): The methods, equipment, facilities, and controls for producing processed food. As the minimum sanitary and processing requirements for producing safe and wholesome food, they are an important part of regulatory control over the safety of the nation's food supply. FDA.

Closed-loop MRP: A system built around material requirements planning that includes the additional planning processes of production planning (sales and operations planning), master production scheduling, and capacity requirements planning. Once this planning phase is complete and the plans have been accepted as realistic and attainable, the execution processes come into play. These processes include the manufacturing control processes of input-output (capacity) measurement, detailed scheduling and dispatching, as well as anticipated delay reports from both the plant and suppliers, supplier scheduling, and so on. The term *closed loop* implies not only that each of these processes is included in the overall system, but also that feedback is provided by the execution processes so that the planning can be kept valid at all times. APICS.

C of A (certificate of analysis): A supplier's certification that the supplies or services in question meet specified requirements. APICS.

Competitive benchmarking: Benchmarking a product or service against competitors. *Synonym:* Performance benchmarking. *See* benchmarking. APICS.

Configurator: Software system that creates, uses, and maintains product models that allow complete definition of all possible product options and variations with a minimum of data entries. APICS.

Conflict mineral supply chain: The supply chain of minerals mined in conditions of armed conflict and human rights abuses, and passed through a variety of intermediaries before being purchased by manufacturing companies. Understanding the origin of these minerals and avoiding procurement of them from countries known to engage in armed conflict, human rights abuses are considered a corporate social responsibility, and many sovereign nations have, or are in the process of passing, laws prohibiting the procurement of conflict minerals.

Continuous improvement (CI): (1) A philosophy of frequently reviewing processes, identifying opportunities for improvement, and implementing changes to get closer to perfection. Osterling. (2) The act of making incremental, regular improvements and upgrades to a process or product in the search for excellence. APICS.

Contract manufacturing: Manufacturers that manufacture a product that has been outsourced to them by their customer based on their specialized core competence and organizational focus. An example would be flavors that enhance a basic recipe/formulation to meet desired taste, color, and nutritional values. The food supply-chain community is extensively populated with contract manufacturers.

Control group cycle counting: A technique of cycle counting focused on the identification of the sources of data errors, and formulating appropriate strategies to remedy the root causes uncovered. Control group cycle counting is the front-end to implementing an effect cycle counting process.

CPIM (certified production and inventory management): The APICS certification that is a recognition of a high level of professional knowledge. APICS.

CRP (capacity requirements planning): The function of establishing, measuring, and adjusting limits or levels of capacity. The term *capacity requirements planning* in this context refers to the process of determining in detail the amount of labor and machine resources required to accomplish the tasks of production. Open shop orders and planned orders in the MRP system are input to CRP, which through the use of parts routings and time standards translates these orders into hours of work by work center by time period. Even though rough-cut capacity planning may indicate that sufficient capacity exists to execute the MPS, CRP may show that capacity is insufficient during specific time periods. APICS.

CSCP (certified supply chain professional): A certification developed and managed by the professional association APICS that covers the design, planning, and controlling of end-to-end domestic and global supply chains. CSCP certification covers the best practices within the walls of each supply-chain partner, and the areas of customer relations, international trade, information technology enablement, and physical logistics. It promotes a common understanding of the vocabulary, resources, and frameworks used within a company to effectively manage a supply chain.

Direct water footprint: The direct water footprint of a consumer or producer (or a group of consumers or producers) refers to the freshwater consumption and pollution that is associated with the water use by the consumer or producer. It is distinct from the indirect water footprint, which refers to the water consumption and pollution that can be associated with the production of the goods and services consumed by the consumer or the inputs used by the producer. Water Footprint Network.

DSHEA (Dietary Supplement Health and Education Act): Congress defined the term *dietary supplement* in the Dietary Supplement Health and Education Act (DSHEA) of 1994. A dietary supplement is a product taken by mouth that contains a dietary ingredient intended to supplement the diet. The dietary ingredients in these products may include vitamins, minerals, herbs or other botanicals, amino acids, and substances such as enzymes, organ tissues, glandulars, and metabolites. DSHEA places dietary supplements in a special category under the general umbrella of foods, not drugs, and requires that every supplement be labeled a dietary supplement.

EFSA (European Food Safety Authority): Responsible for European Union (EU) risk assessment regarding food and feed safety. In close collaboration with national authorities and in open consultation with its stakeholders, EFSA provides independent scientific advice and clear communication on existing and emerging risks. EFSA.

e-Kanban: An electronic signal sent over the Internet to authorize a replenishment activity. e-Kanban replaces the original paper-based Kanban system, and has the advantage of being electronically transmitted to a re-supplying resource internally or externally. A key capability of e-Kanban is the time-based management of the replenishment request, and fulfillment activity through a workflow that has critical scheduled milestones, and sends alerts to all involved when these have not been complied with.

ERP (Enterprise Resource Planning): Framework for organizing, defining, and standardizing the business processes necessary to effectively plan and control an organization so the organization can use its internal knowledge to seek external advantage. APICS.

Functional foods: Functional foods are foods generally considered to offer additional benefits that may reduce the risk of disease or promote optimal health. Examples of so-called functional foods include calcium-fortified orange juice and spreads with omega-3 added. Mayo Clinic.

GHG Protocol: The most widely used international GHG emissions accounting tool for government and business leaders to understand, quantify, and manage greenhouse gas emissions. GHG Protocol.

GHG Scope 1: All direct GHG emissions. Direct GHG emissions are emissions from sources that are owned or controlled by the reporting entity. GHG Protocol.

GHG Scope 2: Indirect GHG emissions from consumption of purchased electricity, heat, or steam. GHG Protocol.

GHG Scope 3: Other indirect emissions, such as the extraction and production of purchased materials and fuels, transport-related activities in vehicles not owned or controlled by the reporting entity, electricity-related activities (e.g., T&D losses) not covered in Scope 2, outsourced activities, waste disposal, etc. GHG Protocol.

GRAS (generally recognized as safe): Under Sections 201(s) and 409 of the Federal Food, Drug, and Cosmetic Act, any substance that is intentionally added to food is a food additive, which is subject to premarket review and approval by FDA, unless the substance is generally recognized, among qualified experts, as having been adequately shown to be safe under the conditions of its intended use, or unless the use of the substance is otherwise excluded from the definition of a food additive. U.S. FDA (see the U.S. FDA website for an expanded definition).

Green water: The precipitation on land that does not run off or recharge the groundwater but is stored in the soil or temporarily stays on top of the soil or vegetation. Eventually, this part of precipitation evaporates or transpires through plants. Green water can be made productive for crop growth (but not all green water can be taken up by crops, because there will always be evaporation from the soil and because not all periods of the year or areas are suitable for crop growth). Water Footprint Network.

Green water footprint: Volume of rainwater consumed during the production process. This is particularly relevant for agricultural and forestry products (products based on crops or wood), where it refers to the total rainwater evapotranspiration (from fields and plantations) plus the water incorporated into the harvested crop or wood. Water Footprint Network.

Gray water footprint: The gray water footprint of a product is an indicator of freshwater pollution that can be associated with the production of a product over its full supply chain. It is defined as the volume of freshwater that is required to assimilate the load of pollutants based on existing ambient water quality standards. It is calculated as the volume of water that is required to dilute pollutants to such an extent that the quality of the water remains above agreed water quality standards. Water Footprint Network.

HACCP (Hazard Analysis Critical Control Point): HACCP is a management system in which food safety is addressed through the analysis and control of biological, chemical, and physical hazards from raw material production, to procurement and handling, to manufacturing, distribution, and consumption of the finished product. FDA.

IE (industrial engineering): The engineering discipline concerned with facilities layout, methods measurement and improvement, statistical quality control, job design and evaluation, and the use of management sciences to solve business problems. APICS.

IFT (Institute of Food Technologists): Exists to advance the science of food. Its long-range vision is to ensure a safe and abundant food supply contributing to healthier people everywhere. IFT website.

Indirect water footprint (virtual): The indirect water footprint of a consumer or producer refers to the freshwater consumption and pollution behind products being consumed or produced. It is equal to the sum of the water footprints of all products consumed by the consumer or of all (nonwater) inputs used by the producer. Water Footprint Network.

Information technology (IT): The technology of computers, telecommunications, and other devices that integrate data, equipment, personnel, and problem-solving methods in planning and controlling business activities. Information technology provides the means for collecting, storing, encoding, processing, analyzing, transmitting, receiving, and printing text, audio, or video information. APICS.

Inventory turns: The number of times that an inventory cycles, or turns over, during the year. A frequently used method to compute inventory turnover is to divide the average inventory level into the annual cost of sales. For example, an average inventory of $3 million divided into an annual cost of sales of $21 million means that inventory turned over seven times. *Synonyms:* Inventory turnover, turnover, inventory velocity. APICS.

ISO (International Organization for Standardization): A group of cooperating institutes from 155 countries working to develop and publish international standards. It acts as a bridge between public and private sectors. APICS.

ISO 22000: Describes the requirements of a food safety management system and incorporates and strengthens the Hazard Analysis Critical Control Point system to create an effective food safety management system. ASQ.

Item master: The collection of data within ERP required to manage an item that makes it unique; includes item classification, current configuration level, cost summary, planning rules, Kosher status, allergen/non-allergen, who is responsible for inventory planning, etc. In some environments, it is referred to as master data.

JIT (just in time): A philosophy of manufacturing based on planned elimination of all waste and on continuous improvement of productivity. It encompasses the successful execution of all manufacturing activities required to produce a final product, from design engineering to delivery, and includes all stages of conversion from raw material onward. The primary elements of just in time are to have only the required inventory when needed; to improve quality to zero defects; to reduce lead times by reducing setup times, queue lengths, and lot sizes; to incrementally revise the operations themselves; and to accomplish these activities at minimum cost. In the broad sense, it applies to all forms of manufacturing—job shop, process, and repetitive—and to many service industries as well. APICS.

Kaizen: (1) An improvement philosophy in which continuous incremental improvement occurs over a sustained period of time, creating more value and less waste, resulting in increased speed, lower costs, and improved quality. When applied to a business enterprise, it refers to ongoing improvement involving the entire workforce, including senior leadership, middle management, and frontline workers. Kaizen is also a philosophy that assumes that our way of life (working, social, or personal) deserves to be constantly improved. Osterling. (2) The Japanese term for improvement; continuing improvement involving everyone—managers and workers. In manufacturing, kaizen relates to finding and eliminating waste in machinery, labor, or production methods. *See* continuous process improvement. APICS.

Kaizen blitz: *See* kaizen event.

Kaizen event: A structured, team-based, problem-solving activity of short duration used to improve processes throughout an organization. Activities typically include: (1) team training, (2) current-state analysis, (3) future-state design, (4) prioritization of improvements, (5) training on a new process, and (6) implementing the selected improvements. Duration is typically one to five days. The event team is focused on the process 100 percent of the time during the event and is cross-functional in composition. Also referred to as kaizen blitz and rapid improvement event (RIE). Osterling.

Kanban: (1) A type of pull production system whereby the downstream process signals the upstream process to replenish what has been consumed. Kanbans typically pull by part number. Kanban means signboard in Japanese. Compare to FIFO lanes. Osterling. (2) A method of just-in-time production that uses standard containers or lot sizes with a single card attached to each. It is a pull system in which work centers signal with a card that they wish to withdraw parts from feeding operations or suppliers. The Japanese word *kanban*, loosely translated, means card, billboard, or sign, but other signaling devices such as colored golf balls have also been used. The term is often used synonymously for the specific scheduling system developed and used by the Toyota Corporation in Japan. APICS.

Kosher (food): Food prepared in accordance with Jewish dietary laws. Jewish dietary laws are rules concerning food that are derived from biblical laws and rabbinical extensions. A person who follows Jewish dietary laws is said to be "keeping kosher."

Lean management system: (1) The philosophy of aggressive, continuous improvement executed through defining value from the customer's perspective, mapping the value streams, creating flow, working at the pull of the customer, and pursuing perfection. Osterling. (2) Lean production—a philosophy of production that emphasizes the minimization of the amount of all the resources (including time) used in the various activities of the enterprise. It involves identifying and eliminating non-value-adding activities in design, production, supply-chain management, and dealing with customers. Lean producers employ teams of multiskilled workers at all levels of the organization and use highly flexible, increasingly automated machines to produce volumes of products in potentially enormous variety. It contains a set of principles and practices to reduce cost through the relentless removal of waste and through the simplification of all manufacturing and support processes. APICS.

Lean manufacturing: *See* Lean management system.

Lot traceability: The ability to identify the lot or batch number of product in terms of one or all of the following: its composition, purchased parts, manufacturing date, or shipped items. In certain regulated industries, lot traceability may be a legislative requirement. APICS.

Maslow's hierarchy of needs: A theory that human needs are arranged in an order, or hierarchy, of importance. The hierarchy includes physiological, safety, social, esteem, and self-actualization needs. APICS.

ME (manufacturing engineering): The engineering discipline concerned with designing and improving production processes. *See* process engineering. APICS.

MES (manufacturing execution system): Programs and systems that participate in shop floor control, including programmed logic controllers and process control computers for direct and supervisory control of manufacturing equipment; process information systems that gather historical performance information, then generate reports; graphical displays; and alarms that inform operations personnel what is going on in the plant currently and a very short history into the past. Quality control information is also gathered, and a laboratory information management system may be part of this configuration to tie process conditions to the quality data that are generated. Thereby, cause-and-effect relationships can be determined. The quality data at times affect the control parameters that are used to meet product specifications either dynamically or offline. APICS.

Mistake-proofing: A device or procedure designed to prevent the generation of defects. The English translations for this Japanese phrase are: *poka*, which means "error," and *yoke*, which means "to avoid." Also referred to as poka-yoke. Osterling.

MPS (master production scheduling): The master production schedule is a line on the master schedule grid that reflects the anticipated build schedule for those items assigned to the master scheduler. The master scheduler maintains this schedule, and in turn, it becomes a set of planning numbers that drives material requirements planning. It represents what the company plans to produce expressed in specific configurations, quantities, and dates. The master production schedule is not a sales item forecast that represents a statement of demand. The master production schedule must take into account the forecast, the production plan, and other important considerations, such as backlog, availability of material, availability of capacity, and management policies and goals. APICS.

MRP (material requirements planning): A set of techniques that uses bill of material data, inventory data, and the master production schedule to calculate requirements for materials. It makes recommendations to release replenishment orders for material. Further, because it is time phased, it makes recommendations to reschedule open orders when due dates and need dates are not in phase. Time-phased MRP begins with the items listed on the MPS and determines (1) the quantity of all components and materials required to fabricate those items and (2) the date that the components and material are required. Time-phased MRP is accomplished by exploding the bill of material, adjusting for inventory quantities on hand or on order, and offsetting the net requirements by the appropriate lead times. APICS.

MRP2 (manufacturing resource planning 2): A method for the effective planning of all resources of a manufacturing company. Ideally, it addresses operational planning in units, financial planning in dollars, and has a simulation capability to answer what if questions. It is made up of a variety of processes, each linked together: business planning, production planning (sales and operations planning), master production scheduling, material requirements planning, capacity requirements planning, and the execution support systems for capacity and material. Output from these systems is integrated with financial reports such as the business plan, purchase commitment report, shipping budget, and inventory projections in dollars. Manufacturing resource planning is a direct outgrowth and extension of closed-loop MRP. APICS.

Nanotechnology: Create, explore, and manipulate materials measured in nanometers (billionths of a meter). Such materials can have chemical, physical, and biological properties that differ from those of their larger counterparts. FDA.

NSF International: An independent, not-for-profit organization that provides standards development, product certification, auditing, education, and risk management for public health and the environment. NSF.

Nutritional supplements: A dietary supplement is a product taken by mouth that contains a dietary ingredient intended to supplement the diet. The dietary ingredients in these products may include vitamins, minerals, herbs or other botanicals, amino acids, and substances such as enzymes, organ tissues, glandulars, and metabolites. FDA.

Operations management: (1) The planning, scheduling, and control of the activities that transform inputs into finished goods and services. (2) A field of study that focuses on the effective planning, scheduling, use, and control of a manufacturing or service organization through the study of concepts such as design engineering, industrial engineering, management information systems, quality management, production management, inventory management, accounting, and other functions as they affect the operation. APICS.

Organic (food): A food product that uses ingredients grown or, in the case of food, livestock fed organic grain or equivalents (free range) using an ecological production management system that promotes and enhances biodiversity, biological cycles, and soil biological activity. It is based on minimal use of off-farm inputs and on management practices that restore, maintain, and enhance ecological harmony. Organic is synonymous with pesticide-free production. Organic food products have an additional layer of regulatory compliance.

Outcome-driven innovation (ODI): A strategy and innovation process developed by Anthony Ulwick. It is built around the theory that people buy products and services to get jobs done. It links a company's value creation activities to customer-defined metrics.

PAC (production activity control): The function of routing and dispatching the work to be accomplished through the production facility and performing supplier control. PAC encompasses the principles, approaches, and techniques needed to schedule, control, measure, and evaluate the effectiveness of production operations. *See* shop floor control. APICS.

Pareto principle (law): A concept developed by Vilfredo Pareto, an Italian economist, that states that a small percentage of a group accounts for the largest fraction of the impact, value, and so on. In an ABC classification, for example, twenty percent of the inventory items may constitute eighty percent of the inventory value. *See* ABC classification, 80/20. APICS.

PDCA (plan-do-check-act): (1) The basic steps to be followed in making continuous incremental improvements (kaizen), adapted by W. Edwards Deming from Walter Shewhart's PDSA cycle (S = study). Osterling. (2) A four-step process for quality improvement. In the first step (plan), a plan to effect improvement is developed. In the second step (do), the plan is carried out, preferably on a small scale. In the third step (check), the effects of the plan are observed. In the last step (action), the results are studied to determine what was learned and what can be predicted. The plan-do-check-act cycle is sometimes referred to as the Shewhart cycle (because Walter A. Shewhart discussed the concept in his book *Statistical Method from the Viewpoint of Quality Control*) and as the Deming circle (because W. Edwards Deming introduced the concept in Japan; the Japanese subsequently called it the Deming circle). APICS.

Planning BOM (bill of material): An artificial grouping of items or events in bill-of-material format used to facilitate master scheduling and material planning. It may include the historical average of demand expressed as a percentage of total demand for all options within a feature or for a specific end item within a product family and is used as the quantity in the planning bill of material. APICS.

Process batch ERP: An enhanced version of ERP that supports the manufacture of formula or recipe-based products such as food, nutritional supplements, pharmaceuticals, and specialty chemicals made in batches. The additional functionality includes subsystems to support regulatory compliance requirements such as lot traceability and HACCP. Another example of the differences between traditional ERP and batch process ERP is the ability to link ingredients and their nutritional or chemical values in a formulation.

Quality functional deployment (QFD): A methodology designed to ensure that all the major requirements of the customer are identified and subsequently met or exceeded through the resulting product design process and the design and operation of the supporting production management system. QFD can be viewed as a set of communication and translation tools. QFD tries to eliminate the gap between what the customer wants in a new product and what the product is capable of delivering. QFD often leads to a clear identification of the major requirements of the customers. These expectations are referred to as the voice of the customer (VOC). APICS.

RCCP (rough-cut capacity planning): The process of converting the master production schedule into requirements for key resources, often including labor, machinery, warehouse space, suppliers' capabilities, and in some cases, money. Comparison to available or demonstrated capacity is usually done for each key resource. This comparison assists the master scheduler in establishing a feasible master production schedule. Three approaches to performing RCCP are the bill of labor (resources, capacity) approach, the capacity planning using overall factors approach, and the resource profile approach. *See* bill of resources, capacity planning, capacity planning using overall factors, product load profile, resource profile. APICS.

Recycling: Using waste as material to manufacture a new product. Recycling involves altering the physical form of an object or material and making a new object from the altered material. Note: Reuse is often confused with recycling. A definition for reuse:

Using an object or material again, either for its original purpose or for a similar purpose, without significantly altering the physical form of the object or material.

Relational database management system (RDBMS): A type of database management system (DBMS) that stores data in the form of related tables. Relational databases are powerful because they require few assumptions about how data are related or how they will be extracted from the database. As a result, the same database can be viewed in many different ways.

ROI (return on investment): A relative measure of financial performance that provides a means for comparing various investments by calculating the profits returned during a specified time period. In the theory of constraints, ROI is calculated as throughput minus operating expense divided by investment. APICS.

SCOR (Supply Chain Operations Reference model): A process reference model developed and endorsed by the Supply Chain Council as the cross-industry, standard diagnostic tool for supply-chain management. The SCOR model describes the business activities associated with satisfying a customer's demand, which include plan, source, make, deliver, and return. Use of the model includes analyzing the current state of a company's processes and goals, quantifying operational performance, and comparing company performance to benchmark data. SCOR has developed a set of metrics for supply chain performance, and Supply Chain Council members have formed industry groups to collect best practices information that companies can use to evaluate their supply chain performance. APICS.

SFC (shop floor control): A system for using data from the shop floor to maintain and communicate status information on shop orders (manufacturing orders) and on work centers. Shop floor control can use order control or flow control to monitor material movement through the facility. The major subfunctions of shop floor control are (1) assigning priority of each shop order, (2) maintaining work-in-process quantity information, (3) conveying shop order status information to the office, (4) providing actual output data for capacity control purposes, (5) providing quantity by location by shop order for work-in-process inventory and accounting purposes, and (6) providing measurement of efficiency, utilization, and productivity of the workforce and machines. The major subfunctions for flow control are based primarily on production rates and feeding work into production to meet these planned rates, then monitoring and controlling production. APICS.

Single-minute exchange of die (SMED): The concept of setup times of less than ten minutes, developed by Shigeo Shingo in 1970 at Toyota. APICS.

Six Sigma: A methodology that furnishes tools for the improvement of business processes. The intent is to decrease process variation and improve product quality. APICS.

Software as a service (SaaS): A software distribution model where an application is hosted as a service provided to customers across the Internet. APICS.

S&OP (sales and operations planning): A process to develop tactical plans that provide management the ability to strategically direct its businesses to achieve competitive advantage on a continuous basis by integrating customer-focused marketing plans for new and existing products with the management of the supply chain. The process brings together all the plans for the business (sales, marketing, development, manufacturing, sourcing, and financial) into one integrated set of plans. It is performed at least once a month and is reviewed by management at an aggregate (product family) level. The process must reconcile all supply, demand, and new product plans at both the detail and aggregate levels and tie to the business plan. It is the definitive statement of the company's plans for the near to intermediate term, covering a horizon sufficient to plan for resources and to support the annual business planning process. Executed properly, the sales and operation planning process links the strategic plans for the business with its execution and reviews performance measurements for continuous improvement. *See* aggregate planning, production plan, production planning, sales plan, tactical planning. APICS.

SOP (standard operating procedure): Written procedure prescribed for repetitive use as a practice, in accordance with agreed upon specifications aimed at obtaining a desired outcome.

Spaghetti diagram: A diagram representing the physical path taken by a product (or service) as it travels through all the steps required to transform a requirement into a deliverable. This can also be used to draw the path walked by those involved in completing the required activities to deliver the product (or service). The diagram derives its name from the way it commonly looks after mapping a process within a function-based organization because the diagram looks much like a plate of spaghetti. Osterling.

SSOP (sanitary standard operating procedure): A written procedure that an establishment develops and implements to prevent direct contamination or adulteration of a product. USDA.

Supply chain: The global network used to deliver products and services from raw materials to end customers through an engineered flow of information, physical distribution, and cash. APICS.

Supply Chain Council (SCC): A global nonprofit organization that developed a comprehensive set of tools to make dramatic and rapid improvements in supply chain processes. This integrated set of tools is called the Supply Chain Operations Reference (SCOR) model, and is the most widely accepted framework for evaluating and comparing supply chain activities and their performance.

Supply-chain management: The global network used to deliver products and services from raw materials to end customers through an engineered flow of information, physical distribution, and cash. APICS.

Sustainability: The ability of an organization to "meet the needs of the present without compromising the ability of future generations to meet their own needs" (World Commission on Environment and Development, *Our Common Future*, 1987). It is generally considered to consist of three dimensions: environmental (the natural environment), social (effect on society), and economic (financial viability). APICS.

Takt time: The pace at which work must be completed to meet customer demand. To calculate, divide the available work time by the customer demand for that period. For example, if a call center receives nine hundred calls per shift, and there are twenty-seven thousand seconds of available work time, the takt time is thirty seconds per call. Therefore, one call must be completed every thirty seconds to meet customer demand. Takt, a German word meaning "pace," is the heartbeat of any Lean system. Process time divided by Takt time yields the number of workers required to support a specific product. Osterling.

Total landed cost: The total cost of a landed shipment, including purchase price, freight, insurance, and other costs up to the port of destination. In some instances, it may also include the customs, duties, and other taxes levied on the shipment.

Total quality management (TQM): A term coined to describe Japanese-style management approaches to quality improvement. Since then, TQM has taken on many meanings. Simply put, TQM is a management approach to long-term success through customer satisfaction. TQM is based on the participation of all members of an organization in improving processes, goods, services, and the culture in which they work. The methods for implementing this approach are found in teachings of such quality leaders as Philip B. Crosby, W. Edwards Deming, Armand V. Feigenbaum, Kaoru Ishikawa, J.M. Juran, and Genichi Taguchi. APICS.

Two bin: A type of fixed-order system in which inventory is carried in two bins. A replenishment quantity is ordered when the first bin (working) is empty. During the replenishment lead time, material is used from the second bin. When the material is received, the second bin (which contains a quantity to cover demand during lead time plus some safety stock) is refilled and the excess is put into the working bin. At this time, stock is drawn from the first bin until it is again exhausted. This term is also used loosely to describe any fixed-order system even when physical bins do not exist. APICS.

UN Global Compact: A strategic policy initiative for businesses that are committed to aligning their operations and strategies with ten universally accepted principles in the areas of human rights, labor, environment, and anticorruption. UN.

Upstream suppliers: A relative reference within a firm or supply chain to indicate moving in the direction of the raw material supplier. APICS.

USDA (U.S. Department of Agriculture): The U.S. Department of Agriculture is the central federal department that works in partnership with state, county, municipal, and tribal governments to use a range of tools to protect the consuming public and farmers and ranchers.

U.S. Food Safety Modernization Act (S.510): The FDA Food Safety Modernization Act (FSMA) is the most sweeping reform of U.S. food safety laws in more than seventy years. It aims to ensure the U.S. food supply is safe by shifting the focus from responding to contamination to preventing it. FDA.

Value stream mapping (VSM): A high-level, visual representation of all of the process steps (both VA and NVA) required to transform a customer requirement into a delivered good or service. A VSM shows the connection between information flow and product flow, as well as the major process blocks and barriers to flow. Value stream maps are used to document current-state conditions as well as design a future state. One of the key objectives of value stream mapping is to identify non-value-adding activities for elimination. Value stream maps, along with the value stream implementation plan, are strategic tools used to help identify, prioritize, and communicate continuous improvement activities. Osterling.

Virtual water content: The virtual water content of a product is the freshwater embodied in the product, not in a real sense, but in a virtual sense. It refers to the volume of water consumed or polluted for producing the product, measured over its full production chain. If a nation exports/imports such a product, it exports/imports water in virtual form. The virtual water content of a product is the same as the water footprint of a product, but the former refers to the water volume embodied in the product alone, while the latter refers to that volume, but also to which sort of water is being used and to when and where that water is being used. The water footprint of a product is thus a multidimensional indicator, whereas virtual water content refers to a volume alone. Water Footprint Network.

Visual workplace (management): An approach to managing products, people, and processes using low-cost, easy-to-understand visual devices. These devices, when properly utilized, will quickly and effectively communicate objectives, performance, operating conditions, and problems. Osterling.

Voice of the customer—QFD (quality functional deployment): Actual customer descriptions in words for the functions and features customers desire for goods and services. In the strict definition, as relates QFD, the term *customer* indicates the external customer of the supplying entity. APICS.

Water footprint: The water footprint is an indicator of freshwater use that looks at both direct and indirect water use of a consumer or producer. The water footprint of an individual, community, or business is defined as the total volume of freshwater that is used to produce the goods and services consumed by the individual or community or produced by the business. Water use is measured in terms of water volumes consumed (evaporated) or polluted per unit of time. A water footprint can be calculated for a particular product, for any well-defined group of consumers (e.g., an individual, family, village, city, province, state, or nation) or producers (e.g., a public organization, private enterprise, or economic sector). The water footprint is a geographically explicit indicator, not only showing volumes of water use and pollution, but also the locations. Water Footprint Network.

Water Footprint Network: A nongovernment organization dedicated to promoting the transition toward sustainable, fair, and efficient use of freshwater resources worldwide by (1) Advancing the concept of the water footprint, a spatially and temporally explicit indicator of direct and indirect water use of consumers and producers; (2) Increasing the water footprint awareness of communities, government bodies, and businesses and their understanding of how consumption of goods and services and production chains relate to water use and impacts on freshwater systems, and (3) Encouraging forms of water governance that reduce the negative ecological and social impacts of the water footprints of communities, countries, and businesses. Water Footprint Network. Please go to the Water Footprint Network for an expanded definition.

Water Footprint Network: A dynamic, international learning community that is a platform for connecting diverse communities interested in sustainability, equitability, and efficiency of water use. Water Footprint Network.

Webinar: Short for web-based seminar, a presentation, lecture, workshop, or seminar that is transmitted over the web. A key feature of a webinar is its interactive elements—the ability to give, receive, and discuss information. Contrast with webcast, in which the data transmission is one way and does not allow interaction between the presenter and the audience.

WHO (World Health Organization): The directing and coordinating authority for health within the United Nations system. It is responsible for providing leadership on global health matters, shaping the health research agenda, setting norms and standards, articulating evidence-based policy options, providing technical support to countries, and monitoring and assessing health trends.

WIP (work in progress): (1) Work that has been made available to be worked on, has been initiated, or has been completed but has not yet been released to the downstream customer. Osterling. (2) A good or goods in various stages of completion throughout the plant, including all material from raw material that has been released for initial processing up to completely processed material awaiting final inspection and acceptance as finished goods inventory. Many accounting systems also include the value of semifinished stock and components in this category. APICS.

Wireless technology: Transmitting signals over invisible radio waves instead of wires. Wireless technologies are used for things as simple as making a phone call or as complex as letting the sales force get information from an ERP application. *CIO Magazine.*

WMS (warehouse management system): A system that manages all processes that a warehouse carries out. These processes include receiving, picking, and shipping. APICS.

Workflow: The sequence of value-adding activities that concludes with the delivery of a product or service, often across different organizational boundaries.

World Wide Web (WWW): A set of software, protocols, hypertext conventions, and multimedia techniques that enable use of the Internet. APICS.

QUIZ ANSWERS

Chapter 1

Question	True	False	Explanation
1.1. This book will benefit only the manufacturer of food products within a specific supply chain.		X	This book covers the entire food ecosystem.
1.2. This book will benefit the manufacturers of food products or nutraceuticals irrespective of the supply chain they participate in.	X		This book describes best operating practices, how they connect to and support regulatory compliance, and how to implement them irrespective of the food or nutra product.
1.3. This book provides no value to the brand owner who utilizes contract manufacturers.		X	This book is a tool for the brand owner to in effect coach its contract manufacturers on how to improve business processes as well as process reliability, which then reduces brand owner liabilities.
1.4. The manufacturer of food and nutra products is a process flow requiring the participation of many departments and disciplines to achieve food safety regulatory compliance, operational efficiency, and superior supply-chain management.	X		A manufacturing company is an integrated set of process flows and organizations that understand this and act accordingly to perform much better than those adhering to old-school, foxhole departmental thinking.
1.5. The targeted reader for this book is restricted to an organization's CEO.		X	This book is written with the knowledge needs of a supervisor, manager, senior executive, and CEO in mind.
1.6. Supply chains will become less complex because of consumer preferences.		X	Highly unlikely; the consumer seems to want more variety.

Question	True	False	Explanation
1.7. There will be a reduction in food safety compliance requirements in the future.		X	Highly unlikely; there always seems to be new or revised food safety legislation pending in the U.S. Congress and other sovereign nations.
1.8. The discipline required for product safety in terms of accurate record keeping and data reporting accuracy is an economic burden with no payback.		X	Accurate record keeping provides the foundation for superior inventory management. Inventory is still the largest expense for most companies, and excessive inventory has a true carrying cost exceeding twenty-eight percent.
1.9. The following chapters of this book are intended to provide guidance on leveraging food safety excellence into superior business operations.	X		That is the mission this book was written to support.
1.10. There is a high probability that the U.S. healthcare reform will lead to changes in the product portfolio and product labeling.	X		With obesity at record levels in the United States and the deaths related to it increasing along with healthcare costs, the attention to foods consumed will increase. For example, if you look at a soup product today you often will have the original version, a low-calorie version, a low-fat version, and a reduced sodium version.
1.11. Future energy costs for those who heat, cold store, or freeze products will not be a major consideration.		X	Energy costs might just be the second biggest cost after inventory in the future. Cheap energy no longer exists in most geographic regions. In recognition of this, some electric utilities are providing funding to manufacturers to implement Lean techniques to produce more product in less time, hence using less energy per unit.

Chapter 2

Luncheon menu exercise

The possible end item combinations are 57,024

(16 protein × 6 bread × 6 condiments × 9 salads × 11 beverages).

Question	True	False	Explanation
2.1. Estimates from the U.S. FDA are that the total number of food facilities that must register with FDA is approximately 100,000.		X	The U.S. FDA estimates over 400,000.
2.2. The food safety regulatory compliance treatment by food safety regulators is uniform across all food and nutritional supplement manufacturing.		X	It is based in part on the characteristics of the food ingredient or combination of ingredients and their interaction with each other and the known impact on food safety.
2.3. Current consumer demands for more variety and dietary considerations are increasing product portfolios and complexity.	X		Go shopping and start counting!
2.4. There is no authoritative text on understanding product modularity and simplifying portfolio management.		X	Please read Joseph Pine's book on mass customization. *Mass Customization* is listed in this appendix.
2.5. Supply chains have become more global.	X		Since the end of the Cold War global trade has substantially increased. The PRC has the third largest export share of food to the United States.
2.6. Fuels used in international trade, both ocean and air transport, are taxed.		X	This has distorted the true landed cost of goods imported in many geographies.
2.7. The U.S. FDA in the case of generally recognized as safe (GRAS) submissions has put the burden of proof on manufacturers.	X		This is a burden but also a good example of industry self-regulation.

Question	True	False	Explanation
2.8. Nanotechnology will have no impact on the food and supplements industries.		X	Limited use of nanotechnology has already been deployed, and acceptance varies by culture and region.
2.9. The use of contract manufacturers is not widely used.		X	It is pervasive and consistent with the adoption of the core competence model.
2.10. The effective utilization of information technology for planning, control, data integrity, and record keeping is a constant among small, medium, and large supply chain participants.		X	Many organizations have unconnected "islands of information" that make planning, controlling, and data management difficult.
2.11. Intra-industry benchmarking is common within food and supplement manufacturing.		X	Empirical observation suggests otherwise.
2.12. Food and supplement manufacturing has limited support from educational institutions and forums.		X	There are many universities with food science curriculums and professional associations such as IFT and over two hundred highly focused industry advocacy groups in North America alone.
2.13. There is limited commitment to the United Nations (UN) Global Compact, including the sustainability component.		X	There is substantial support among major corporations, and there is momentum, including a focus on small- to medium-sized enterprises.
2.14. All nations involved in the global food supply chain have demonstrated enthusiasm for the UN Global Compact.		X	The lack of participation by companies from certain countries suggests the need for extraordinary due diligence in selecting suppliers.

Chapter 3

Question	True	False	Explanation
3.1. Food supply chains have been increasingly localized.		X	Since the end of the Cold War food supply chains have become increasing global.
3.2. The challenges in ensuring a safe global food supply chain are minimal.		X	Different food safety approaches and systems representing different views of food and the rapid emergence of a global supply chain have outpaced the development of unified food safety laws, methodologies, and practices.
3.3. In the United States the responsibility for ensuring food safety rests exclusively with the FDA.		X	Responsibilities have been distributed across the: • USDA • EPA • FSIS
3.4. Prior to thc Bioterrorism Act there were no meaningful food safety laws in the United States.		X	There have been numerous food safety laws prior to the Bioterrorism Act. The free enterprise mindset of the United States has always been a factor, and balancing profit and safety a challenge. Major safety incidents moved the safety perspective to the fore and the evolution of food science has helped.
3.5. Lot traceability is a recent concept.		X	Lot traceability had its beginnings in the late 1800s as a way of establishing commercial (financial) liability.
3.6. GMP outline a set of requirements and SOPs how an organization will satisfy them.	X		GMP are the what and SOPs are the how.

Question	True	False	Explanation
3.7. HACCP is designed around the concept of inspecting for defects.		X	HACCP is centered on analysis, science, and designing in preemptive measures and not accepting the result of broken processes.
3.8. The U.S. Bioterrorism Act doesn't address lot traceability.		X	One-up and one-down documentation is required within very short timeframes.
3.9. The United States has been the only country concerned with food safety.		X	The EU and historically Japan have a food safety agenda.
3.10. There are no analytical tools to help prepare for an intentional attack on a food supply chain.		X	The FDA in collaboration has developed or sponsored a series of tools.
3.11. Food and nutritional supplements manufacturers operate under one common GMP.		X	The new nutritional supplements GMP are much more stringent than the food GMP.
3.12. Food safety requirements continue to evolve.	X		Is this a trick question?

Chapter 4

Question	True	False	Explanation
4.1. The purpose of this chapter was only to examine each of the top ten food industry issues.		X	It also described interdependencies each issue had with the others, pointed to the areas that should be addressed first, and provided general guidance on how to resolve the issue.

Question	True	False	Explanation
4.2. Since the passage of the record-keeping requirement for lot traceability in the USA Bioterrorism Act the food industry has been very successful in developing complete and accurate lot trace records.		X	A sample audit by the U.S. HHS inspector general demonstrated that the field-to-fork record-keeping performance (completeness, accuracy) was very poor.
4.3. The record-keeping tools used today for lot traceability are generally the same for both large and small-sized companies.		X	This is a generalization, but larger organizations use process batch ERP, which has lot traceability built in by design. Smaller companies tend to use electronic spreadsheets or keep manual records. There is advantage to the company using process batch ERP effectively.
4.4. Today information technology (IT) systems are cost-effective for large-sized companies only.		X	Most systems and the required equipment have essentially been commoditized and costs are very low, and when properly used have an exceptional ROI.
4.5. Skilled labor is readily available and at low cost.		X	The resource pool of skilled labor available today is very shallow and expensive.
4.6. There is a blueprint and process to developing a skilled and semiskilled workforce.	X		There are ten steps (the ten commandments) that can be implemented with a number of external resources to help a company educate and train a workforce into a high-performance team.
4.7 Improving equipment maintenance can be accomplished by what four approaches or techniques?			1. 6S 2. Simple operator maintenance 3. Preplanned preventive maintenance 4. SMED

Question	True	False	Explanation
4.8. Verbal instructions to a semiskilled worker suffice as adequate employee training.		X	The supervisor is important, but an employee trained to do documented standard work will reduce food safety risk and improve consistency of output.
4.9. There is a direct correlation between employee education and training and maintenance improvements and effective use of automation.	X		Research studies support a yes.
4.10. ERP was designed around the goal of systems integration.	X		Initially ERP's predecessors focused on manufacturing, but as it evolved, other functions were integrated by design, including engineering, QA, sales, procurement, and finance.
4.11. ERP has been enhanced for food and nutritional supplement manufacturing.	X		The version of ERP enhanced to support food and supplements manufacturing is process batch ERP.
4.12. There is a linkage between the philosophy that a company should focus on its core competence and the current supply chain thinking.	X		The adoption of the core competence philosophy meant that more of not only product produced in the factory but also many support processes would be out-sourced.
4.13. There is consensus on what supply-chain management is.		X	There are differing views of what a supply chain is and what is most important. A logistics or distribution or purchasing or manufacturing professional will often have a view and definition centered on his or her role. The broader and more comprehensive perspective can be found with professional associations such as APICS and the Supply Chain Council.

Question	True	False	Explanation
4.14. Supply chains are less complicated today than twenty years ago.		X	For most industries they are much more complicated because of the adoption of the core competence model and the pursuit of global sources of supply with the end of the Cold War.
4.15. Lack of visibility throughout many supply chains is a significant problem.	X		In most all research studies lack of visibility is ranked as a top three issue.
4.16. Most companies have a detailed understanding of the supply chains they participate in, from field to fork.		X	Empirical evidence says no. If they did the Supply Chain Council would have hundreds of thousands of companies as members, but currently only have approximately a thousand member companies spread across many different industries despite a low cost of involvement.
4.17. There are cost-effective resources to help a company improve its systems and supply chain integration based on best practices.	X		The resources available are plentiful, for example: • Supply Chain Council (SCC)—the SCOR model. • APICS—Best practices education in operations and supply-chain management. • Local universities offering operations and supply chain curriculum. • In the United States, the NIST MEP (http://www.nist.gov/mep/)—government-supported consulting assistance to small manufacturers.

Chapter 5

Question	True	False	Explanation
5.1. Education and training on operational best practices, the role each person plays in maintaining food safety, and the effective use of information technology are cost-prohibitive for the small manufacturer.		X	In the past education or training meant expensive classroom instruction. Today, using technologies such as YouTube and the World Wide Web and other inexpensive technologies has allowed professional associations, device suppliers, information technology suppliers, and the manufacturers themselves to put together cost-effective courseware.
5.2. Adhering to food safety regulatory requirements negatively impacts operating efficiency.		X	The discipline, commitment to reliable processes, attention to detail, and pursuit of data accuracy required for effective food safety regulatory compliance mirror the prerequisites for world-class inventory, operations, and supply-chain management, which translates into superior profitability and competitiveness.
5.3. Achieving a lot trace record accuracy of ninety percent is considered acceptable.		X	It is doubtful that any government inspector or customer would accept this low level of accuracy as adequate. The consequence of having this level of accuracy in the event of a recall would mean an expanded recall scope because of the lack of confidence by the authorities or customers in their supplier's record keeping.
5.4. There is no relationship between the physical layout of a work area and accurate record keeping.		X	Any layout that makes it difficult to report at the point where work is performed as it's performed increases the probability of errors.

Question	True	False	Explanation
5.5. Name two techniques that can help in understanding existing workflows and what areas need to be focused on to improve them.			1. Spaghetti diagramming 2. Value stream mapping (VSM)
5.6. What three cost-effective information-based technologies can help reduce data errors?			1. Bar coding 2. Wireless 3. Enterprise Resource Planning (ERP)
5.7. The visual workplace keeps everyone up to date on the status of work in process within a work area.	X		That is the central purpose of the visual workplace.
5.8. The visual work area is difficult to establish.		X	There is trial and error until it's right. Think in terms of "tuning in" a distant radio station on an old-fashioned radio. Successful companies often visit others to see their layouts and creative results.
5.9. What two obstacles need to be removed to set up a visual work area?			1. Limited space (yes it may make sense to go vertical) 2. Overcrowding; this can even occur with unlimited space!
5.10. Keeping lot numbers segregated in a warehouse or stockroom is optional.		X	Not unless you want the FDA threatening to shut your company down, and your customers terminating the sell–buy relationship with your organization!
5.11. The extra space needed to organize work areas to minimize errors is an ongoing perpetual cost.		X	Not necessarily. Companies who have successfully embraced Lean manufacturing often have reduced their square footage footprint because of reduced inventory and quicker processing of product through the factory while retaining all the best aspects of the visual and organized workplace.

Question	True	False	Explanation
5.12. The obligation to maintain data accuracy is limited to receiving, inspection (QA), warehousing, and production staff.		X	It starts with recipe formulation, product processing instructions, and includes every support function.
5.13. All individuals have the aptitude for accurately recording information.		X	Not all individuals have the aptitude to count and record accurately. This isn't a defect but just a manageable constraint. Achieving accuracy can be designed into many processes, work area layouts, and the use of devices to minimize data input, but there will be some situations that require this elusive aptitude.
5.14. The consequences for small data errors are negligible.		X	Small data errors get amplified as they move the product through the booking of a customer order to shipment journey. The consequence of this is loss of confidence in information, and ultimately confidence in food safety.

Chapter 6

Question	True	False	Explanation
6.1. Name three practices emphasized in ERP that are shared by best practices in food safety.			Any three of the following are an appropriate answer: • Creating workforce awareness of cause-and-effect relationships • Understanding process flows • Using proven techniques • Careful planning • Achieving and maintaining data accuracy • Discipline in executing the plan(s) • Free flowing and timely communications • A commitment to professionalism
6.2. ERP is a framework for organizing, defining, and standardizing the business processes necessary to effectively plan and control an organization so the organization can use its internal knowledge to seek external advantage.	X		ERP is the *de facto* combination of systems to plan and control the operations of an enterprise.
6.3. ERP is the result of over fifty years of refinement and evolution.	X		ERP's evolutionary path progressively incorporated inventory planning, inventory control, capacity management, factory scheduling and control, costing and financial ledgers, customer order management, and procurement support.

Question	True	False	Explanation
6.4. In the late 1940s pent-up demand for goods and the lack of systems to balance demand and supply forced organizations to figuratively surround themselves with costly inventory, and in response the search for better inventory planning methods started.	X		Excessive inventory is expensive and estimated to cost over eighteen percent, *excluding* the cost of money.
6.5. Material requirements planning (MRP) was the first step in ERP's evolution and was designed to calculate down through a multilevel recipe (BOM) what inventory was needed to support actual customer demand.	X		The logic of MRP utilized the calculation speed of a computer to calculate: 1. How much was needed 2. When it was needed 3. When it should be started
6.6. What four root causes of success or failure in using MRP by early adopters were uncovered in an in-depth analysis?			1. MRP was "nervous" and frequently changed the provisional raw material plan when demand at the customer order level changed because it was designed for the calculation of dependent demand. Another system was needed to manage customer demand (independent demand) before releasing it to MRP. MPS was designed to do this. 2. Data accuracy was found to be absolutely critical, not unlike the requirements of food safety record keeping. 3. The lack of factory-floor-level detailed scheduling, capacity management, and status monitoring systems. 4. Educating those who would use the system on the best practices.

Question	True	False	Explanation
6.7. What two systems were added to MRP to increase its successful use?			1. Master production scheduling (MPS). MPS was designed to push *decision making* up to the customer order or intermediate (subassembly) level before authorizing MRP to do the numerous detailed calculations to determine raw material requirements. Its role is to stabilize and synchronize the overall corporate material plan. 2. Shop floor scheduling, capacity planning, and monitoring systems.
6.8. What name tag was put on the next evolutionary step that enhanced MRP?			Closed-loop MRP
6.9. With widespread success closed-loop MRP was enhanced and renamed MRP2. List the seven enhancements included in it.			1. Sales and operations planning (S&OP) 2. Customer order management 3. Integration with the financial ledgers 4. Enhanced support of procurement 5. Additional methods of planning (kanban, etc.) 6. Other styles of managing: recipe-based manufacturing, aerospace and defense, etc. 7. Global features: multicurrency, multilanguage, different sovereign taxation requirements, etc.
6.10. Many studies on the benefits of MRP2 were conducted; one is referenced in this chapter. What number of inventory turns improvement was noted in the report?			The Clemson Study showed an improvement from 4.5 to 7.9 turns with an estimated achievement of 11.2.

Question	True	False	Explanation
6.11. What is the cost of excessive inventory, inventory beyond immediate needs (perfect demand-supply balancing)?			Eighteen percent annually *plus* the cost of borrowing money (interest rate). Excessive inventory is expensive!
6.12. MRP2 evolved into ERP, which is the current *de facto* framework to manage a manufacturing enterprise. Most of this evolution was the result of leveraging advanced information technology. Name three of the nine contributors to the evolution of ERP mentioned in this chapter.			Any three of the following are great answers: 1. Improved integration between the various MRP2 subsystems 2. The ability to manage the centralized database more efficiently 3. The ability to rapidly rewrite software instructions to process data and perform analysis quicker 4. Improved human engineering to support decision making 5. Using new emerging tools based on artificial intelligence (AI) 6. Distributing planning and control systems over computer networks both internal and externally 7. Information portals that aggregated both structured (MRP2 outputs) and unstructured data from various sources 8. Using the web as a communications conduit 9. Workflow views of producing a product or service versus the traditional hierarchical views

Question	True	False	Explanation
6.13. Name four different processes or requirements that process batch ERP supports that traditional hard goods ERP doesn't (there are many more).			A partial list of differences: 1. Lot trace, one-up, one-down 2. Extensive regulatory oversight; cGMP, HACCP, third party, etc. 3. Comprehensive ingredient and nutritional labeling 4. Many possible units of measure and unit-to-unit conversions 5. Continuing proof of compliance 6. Shelf life management (plus additional customer-imposed remaining shelf life at time of shipment) 7. Allergen management
6.14. ERP and the Lean management system are complementary.	X		ERP provides a stable framework for interconnected processes (and a centralized data repository) that the Lean management system can be worked within to reduce waste in all forms.
6.15. What concept foretold the extensive out-sourcing that occurred during the last decade?			The concept of operating on an organization's core competence and out-sourcing everything else. The futurist Charles Handy called it the three-leaf-clover organization: • First leaf: The core team of highly focused individuals representing the organization's core competence. • Second leaf: Specialists producing or providing services needed by the organization but not centered on the core competence.

Question	True	False	Explanation
			• Third leaf: Skilled workers brought on board as needed but not assured a permanent position. Mr. Handy's book (*The Age of Unreason*) is recommended reading and listed in this appendix.
6.16. What three developments have led to the rapid development of extended and complex global supply chains?			1. The acceptance of the core competence concept and extensive out-sourcing (see 6.15). 2. The end of the Cold War and the push by major corporations to operate globally. 3. The emergence of the Internet as a global communications and business-to-business transaction backbone.
6.17. The move to extended supply chains hasn't been problem-free; besides risk management, what other opportunity for improvement has surfaced?			Visibility across the supply chain and in particular communications to upstream suppliers. The *Wall Street Journal* article referenced in this book (Clarity Missing in Supply Chain) highlights how destructive lack of visibility can be. In many respects this problem is shared by the current field-to-fork lot traceability dilemma.
6.18. With the move to extensive out-sourcing and the extended supply chain model the importance of ERP diminished.		X	Every organization requires managing the resources within its four walls, including the supply chain (orchestrator) that is closest to the consumer. The expression "a chain is only as strong as its weakest link" applies. An organization (a link) that properly uses ERP is a strong link and reliable supply chain partner.

Question	True	False	Explanation
6.19. Name the four major channels or forums to improve supply-chain management.			1. The SCOR Framework developed by the Supply Chain Council 2. The APICS Certified Supply Chain Professional (CSCP) certification that is supported by courseware taught by qualified instructors 3. The APICS Certified in Production and Inventory Management (CPIM) certification that is supported by courseware taught by qualified instructors. All the processes inherent in process batch ERP are based on the concepts and techniques incorporated into the CPIM curriculum. 4. Universities that offer a certificate program on operations or supply-chain management.

Chapter 7

Question	True	False	Explanation
7.1. What is Lean?			A management system
7.2. What are the four primary objectives of Lean?			1. Deliver maximum value to customers 2. Eliminate all forms of waste 3. Enable the entire workforce and utilize its potential 4. Continuously improve the enterprise

Question	True	False	Explanation
7.3. What is the relationship between Lean and continuous improvement (CI)?			Lean is the management system; CI is an umbrella term that has proven techniques within it to achieve the Lean management system's objectives.
7.4. Implementation of Lean requires extensive investment in capital equipment.		X	Lean/CI is focused on getting more from less by enabling its workforce to continuously improve through collaboration.
7.5. Which technique supports the PDCA cycle: 6S, A3 or Kanban?			A3 supports the PDCA cycle. 6S is an approach to keeping a work area clean, uncluttered, and functional. Kanban is an inventory planning and control technique.
7.6. The A3 method supports visualization of a problem and the proposed resolution.	X		A3 includes visualization but also stresses careful fact gathering, information validation, concise written statements, development of group acceptance, and the need to go through iterative cycles until the best possible proposal is developed.
7.7. What technique systematically surfaces customer needs?			Quality functional deployment (QFD) or outcome-driven innovation (ODI). ODI is considered an alternative to QFD and is a more recent methodology.
7.8. What quality permeates an organization with a Lean culture?			Respect for the individual
7.9. What is the primary purpose of 6S?			6S is focused on eliminating some of the sources of waste through simplification, creation of order, visualization, and making problems easy to see.

Question	True	False	Explanation
7.10. What analytical tool objectively and systematically identifies the status and performance of each node within an end-to-end process?			Value stream mapping (VSM). VSM combines visualization of each connected process step, with current performance data for each, and what information systems support those steps. The current-state VSM identifies which areas should be targeted for kaizen events, and also facilitates the development of the desired future state.
7.11. What is the goal of a kaizen event/blitz?			Make changes within a work area to eliminate one or more forms of waste.
7.12. What are the two methods used to identify potential kaizen event candidates?			1. Creation of a VSM for an end-to-end process 2. Identifying process steps that can potentially cause possible worker injury, or have poor task ergonomics
7.13. Kaizen events are generally spontaneous.		X	A kaizen event/blitz: 1. Is authorized 2. Is a planned activity 3. Is conducted by a team 4. Has significant targeted improvements 5. Has a budget 6. Supports progress toward the future state
7.14. What is the difference between a kaizen event and a kaizen?			1. Kaizen event: planned, structured, big results 2. Kaizen: spontaneous, generally small improvements
7.15. Name a few concepts or techniques used to guide the organization of a work area and its evolution.			1. 6S 2. Manufacturing cell design concepts 3. Visual workplace 4. Kaizen events

Question	True	False	Explanation
7.16. What is the relationship between food safety procedures and standard work?			Both are centered on producing a safe product, following well-designed instructions, executed in a repeatable and disciplined manner.
7.17. Material requirements planning (MRP) is the only proven material planning and control technique available.		X	There are other techniques. Not *all* finished and raw materials should be managed using one technique. The characteristics of an item dictate the choice of technique.

Chapter 8

Question	True	False	Explanation
8.1. The current driver on sustainability is solely due to ecological (green) considerations.		X	The cost of energy generated from fossil fuels has become very expensive and is forecasted to become more costly. GHG emissions and energy are synonymous. Mismanagement of energy will eventually push an organization into an uncompetitive position.
8.2. The U.S. FDA special report *Pathway to Global Product Safety and Quality* forecasts adequate supply of natural raw materials.		X	The report forecasts shortages and increasing global competition. Only those committed to continuous improvement and sustainability will survive in the long term.
8.3. Recycling is the only method to reduce physical landfill waste.		X	Redesign of product and packaging to minimize inputs to landfill or recycling is the long-term preferred approach.

Question	True	False	Explanation
8.4. Products made from recycled materials always use no water or energy resources.		X	Some recycling processes are energy or water intensive. The data to confirm the usage versus virgin are readily available. The ultimate action is redesign to eliminate if possible.
8.5. Early adopters of sustainability measures will be focusing on their supply chains in the near future.	X		Those who made progress have done the analysis, and there is significant opportunity for reduction within the upstream supplier network.
8.6. Currently there are no tools/methodologies to model and manage an end-to-end supply chain.		X	The SCOR modeling methodology (Supply Chain Council) has been developed over a fifteen-year period and is proven and codified. APICS has developed courseware and certification on supply-chain management.
8.7. Water is rapidly becoming a scarce resource within certain geographic regions.	X		The United Nations, in recognition of this critical problem, has started water management initiatives during the last decade. Progress is being made, but the problem is still pressing and awareness and actions need to be accelerated. The food industry is water intensive.
8.8. What is the Water Footprint Network methodology?			It is a comprehensive water accounting system that replaces misleading and antiquated approaches. The combination of it and SCOR (see above) is powerful in managing water both within the four walls of the food processor and across the supply chain.

Question	True	False	Explanation
8.9. What are the three types of water within the Water Footprint Network methodology?			1. Green (rainfall) 2. Blue (catchment/rivers, lake and aquifers) 3. Gray (polluted)
8.10. During the manufacture of a food product what two types of water are involved?			1. Direct—water used by the food processor. 2. Virtual—water inherited in the raw materials used by the food processor. The water accounting process (direct and virtual) is cumulative.
8.11. What global organization has focused attention on water management?			The most prominent is the United Nations with its CEO mandate initiative.
8.12. Reducing energy consumption can only be accomplished through the acquisition of next-generation capital equipment to replace energy-intensive equipment.		X	Corporations have utilized tools in the Lean system to eliminate waste in processes. Thought-leading public utilities have participated in and sponsored Lean projects focused on energy consumption reduction within their customer base.
8.13. What was the UN landmark sustainability initiative that has involved the private sector?			The UN Global Compact (2000). It has a sustainability component drawn from nearly thirty years of UN sponsorship of sustainability measures.
8.14. What two major organizations have responded to the UN initiative for sustainability for GHG emissions management?			1. The Greenhouse Gas Protocol (GHG Protocol), which is more of a consortium 2. The Carbon Disclosure Project (CDP)
8.15. What organization has developed the *de facto* guidance on GHG accounting and reporting?			The Greenhouse Gas Protocol (GHG Protocol)

Question	True	False	Explanation
8.16. Are there standards that support the guidance on GHG accounting and reporting?	X		The International Organization for Standardization (ISO) has created standards in collaboration with the GHG Protocol to support it.
8.17. What are GHG accounting and reporting conceptually similar to those of a publicly traded corporation required to do?			Financial reporting
8.18. What are the three distinct types of emissions generation categories that are contained within the GHG accounting and reporting methodology?			1. Scope 1: Emissions generated within a facility that may be rolled up within an operation boundary and then up to an organizational boundary. 2. Scope 2: Purchased electricity. 3. Scope 3: Emissions generated by upstream suppliers in the manufacture of products for the corporation as well as downstream emissions.
8.19. Which one of the categories is currently considered optional?			Scope 3, but that will change within this decade.
8.20. Is there a requirement for absolute precision in reporting for corporations new to the GHG accounting and reporting processes?		X	Corporations are encouraged to start accounting and reporting even with imperfect data as long as the principles of the methodology are adhered to and there is a commitment to improving emission reduction actions as well as the accounting and reporting process.

Question	True	False	Explanation
8.21. What is the Carbon Disclosure Project (CDP)?			A database of GHG emissions reports from corporations committed to sustainability, often using the GHG Protocol as guidance, to serve business stakeholders such as institutional investors and large corporations. CDP functions as a figurative middle man in channeling requests between different entities.
8.22. Name two major stakeholders who utilize the CDP's capabilities.			1. Large institutional investors managing trillions of USD 2. Major corporations managing all three scopes of their carbon footprint, e.g., Wal-Mart
8.23. What is the name of the UN initiative on water management, and to whom is it addressed?			• UN Global Compact's CEO Water Mandate • The CEO of corporations
8.24. With the finalization of GHG Protocol's Scope 3, what will early adopter large food retailers and large food companies concentrate on?			Their upstream supply chains because they represent both biggest risk and biggest opportunity for GHG *and* cost reductions

WEBSITES

Food Safety

- CARVER + Shock: The FDA site dedicated to this food defense tool, which can be downloaded free of charge and used to understand what is needed at a specific site to be food defense compliant. http://www.fda.gov/ForConsumers/ConsumerUpdates/ucm094560.htm.
- European Common Food Law: The official website covering the European Common Food Law. http://ec.europa.eu/food/food/foodlaw/index_en.htm.
- Biotoxin attack simulation: An organization committed to the development of analytical food safety and defense systems to prevent unintentional or international (terrorist) attacks or, if required, recovery plans. http://www.btsafety.com/index.html.
- European Food Safety Authority (EFSA): http://www.efsa.europa.eu/. Note: EFSA is chartered to provide science-based risk assessment to the EU. FDA: http://www.fda.gov/.
- Food safety history: Significant dates in U.S. food and drug regulatory compliance. http://www.fda.gov/AboutFDA/WhatWeDo/History/Milestones/ucm128305.htm.
- Food Safety Inspection Service (FSIS) for food recalls: http://www.fsis.usda.gov/fsis_recalls/Open_Federal_Cases/index.asp. http://www.fsis.usda.gov/fsis_recalls/Recall_Case_Archive_2010/index.asp. Note: The FSIS operates within the USDA.
- HACCP: The definitive overview of HACCP from the U.S. FDA. http://www.fda.gov/Food/FoodSafety/HazardAnalysisCriticalControlPointsHACCP/HAC.
- HACCP: A definition and overview on HACCP from the University of Florida. http://edis.ifas.ufl.edu/fs122.
- Food imports: USDA e-news letter on food imports. http://www.ers.usda.gov/amberwaves/february08/datafeature/.
- NSF International: A not-for-profit, nongovernmental organization centered on standards development, product certification, education, and risk management for public health and safety. http://www.nsf.org/.
- USDA: http://www.usda.gov/wps/portal/usda/usdahome.
- Nutritional supplements (nutraceuticals): FDA website covering nutritional supplement GMP. http://www.fda.gov/Food/Dietary

Supplements/GuidanceComplianceRegulatoryInformation/RegulationsLaws/ucm110858.htm#fr.

- Report on the sample audit for complete lot trace history by the inspector general of the Human Health Services (HHS) (United States): http://www.hhs.gov/asl/testify/2009/03/t20090326a.html.

Magazine Focused on Operations in the Food Industry

- Food Engineering: http://www.foodengineeringmag.com/.

Emerging Food-Related Technologies

- FDA Nanotechnology Report: http://www.fda.gov/ScienceResearch/SpecialTopics/Nanotechnology/NanotechnologyTaskForceReport2007/default.htm.
- Helmut Kaiser Consultancy Group Study on Nanotechnology: Nanotechnology in Food and Food Processing Industry Worldwide 2008–2010–2015: http://www.hkc22.com/Nanofood.html.
- Functional foods, USDA website on functional foods and antioxidants: http://fnic.nal.usda.gov/nal_display/index.php?info_center=4&tax_level=3&tax_subject=358&topic_id=1610&level3_id=5947&level4_id=0&level5_id=0&placement_default=0.

Professional Associations and Other Organizations That Support Excellence in Operations and Supply-Chain Management

- APICS (Association for Operations and Supply Chain Management): http://www.apics.org.
- AME (American Manufacturing Excellence): http://www.ame.org/. Note: Focused on the Lean enterprise.
- ASQ (American Society for Quality): Food special interest group. http://asq.org/food/membership.html.
- CSCMP (Council of Supply Chain Management Professionals): http://cscmp.org/Default.asp.
- IFT (Institute of Food Technologists): http://www.ift.org/.
- ISM (Institute of Supply Management): http://ism.ws/.
- MESA (Manufacturing Enterprise Solutions Association): http://www.mesa.org/en/index.asp.
- NIST (U.S.): http://www.nist.gov/mep.

- Supply Chain Council: http://supply-chain.org. Note: Developers of the supply chain modeling methodology SCOR.
- Supply chain, GS 1 US: An organization committed to improving standardized supply chain visibility. http://www.gs1us.org/sectors/foodservice.
- Manufacturing Extension Partnership: An organization funded by the U.S. government dedicated to helping the small- to medium-sized manufacturer through cost-effective best practices consulting. http://www.nist.gov/mep/.

Professional Practitioner Certifications

- APICS CPIM certification: http://www.apics.org/certification/cpim/.
- APICS CSCP certification: http://www.apics.org/certification/CSCP/.

Sustainability

- Water supply chain (Water Footprint Network): http://www.waterfootprint.org.
- United Nations publication *Supply Chain Sustainability: A Practical Guide for Continuous Improvement*: http://www.unglobalcompact.org/docs/issues_doc/supply_chain/SupplyChainRep_spread.pdf.
- International Organization for Standardization (ISO) Environmental Standards:
  - Environment management systems: ISO 14004.
  - Greenhouse gas (GHG): ISO 14064:1, http://www.iso.org/iso/home.htm.

- Greenhouse Gas Protocol (GHG): http://www.ghgprotocol.org/.
- World Business Council for Sustainability Development (WBCSD): http://www.wbcsd.org/home.aspx.
- World Resources Institute (WRI): http://www.wri.org/.
- Carbon Disclosure Project: https://www.cdproject.net.
- Case study—application of Lean/continuous improvement techniques to reduce energy consumption: http://www.cmtc.com/cmtc-case-studies/hills-pet-nutrition/.

Miscellaneous

- Food industry information source: Produce Marketing Association (PMA), an industry association providing field-to-fork information. http://fieldtofork.pma.com/?cat=7.
- Import/export related: Fuel tax, no fuel tax on transport of international goods. http://www.nytimes.com/2008/04/26/business/world business/26food.html.
- Online/web-based operations and supply chain training: APICS CPIM and CSCP certification preparation instruction. http://fvtc.edu/public/(Fox Valley Technical College).
- Workforce training best practices: American Society for Training and Development (ASTD). http://www.astd.org/.
- Social media used as a best practices internal or external forum: LinkedIn—A professional web-based community. http://www.linkedin.com/.
- In-house training tool: YouTube—a useful video-based tool accessible via the Internet for customized workforce training. http://www.youtube.com/.
- A non-industry-specific information source on supply chain developments: Supply Chain Brain. http://www.supplychainbrain.com/.

RECOMMENDED READING

Additional cost considerations for carrying surplus inventory: http://www.positive-way.com/business/truecost.htm.

Albrecht, Karl, Zemke, Ron. *Service America!: Doing Business in the New Economy.* Dow-Jones Irwin. 1985 (Re-released by the authors).

APICS Operations Management Body of Knowledge (OMBOK) Framework: http://www.apics.org/resources/ombok.htm.

GHG Protocol: The Green House Gas Protocol: A Corporate Accounting and Reporting Standard, Revised Edition. Available free from the GHG Protocol website.

Moore, Geoffrey A. Crossing the Chasm: Marketing and Selling High-Tech Products to Mainstream Customers. Harper Business Essentials. New York, NY. 2002.

Pine, B. Joseph. *Mass Customization: The New Frontier in Business Competition.* Harvard Business Review Press.

Supply Chain Operations Reference (SCOR) model: http://supply-chain.org/scor.

Recommended reading on the origins of the core competence model:

APICS Dictionary: http://www.apics.org/Resources/APICSDictionary.htm.

Barnard, William and Wallace, Thomas F. *The Innovation Edge—Creating Strategic Breakthroughs Using the Voice of the Customer*. Oliver Wight Publications. *Note:* Focused on QFD. Out of print but available through the used book marketplace.

Carrying cost (of inventory) cost contributors: http://www.remassoc.com/portals/0/rem-precc.pdf.

Handy, Charles. *The Age of Unreason*. Harvard Business School Press. Boston, MA. 1990.

Hoekstra, Arjen Y. *The Water Footprint of Food*, Twente Water Centre, University of Twente, The Netherlands. Available as a free PDF from: http://www.waterfootprint.org/Reports/Hoekstra-2008-WaterfootprintFood.pdf

Hoekstra, Arjen Y., Chapagain, Askok K., Aldaya, Maite M. and Mekonnen, Mesfin M. *The Water Footprint Assessment Manual: Setting the Global Standard*. Routledge, Oxford UK. 2011.

Martin, Karen and Osterling, Mike. *The Kaizen Event Planner: Achieving Rapid Improvement in Office, Service, and Technical Environments*. Productivity Press. New York. 2007.

QFD (Quality Functional Deployment) and Outcome Driven Innovation (ODI)

Rother, Mike and Shook, John. *Learning to See: Value Stream Mapping to Add Value and Eliminate MUDA*. Lean Enterprise Institute. Cambridge, MA. 2003.

Shook, John. *Managing to Learn: Using the A3 Management Process*. Lean Enterprise Institute.

Ulwick, Anthony W. *What Customers Want: Using Outcome-Driven Innovation to Create Breakthrough Products and Services*. McGraw-Hill. New York, NY. 2005

Walker, William T. *Supply Chain Architecture: A Blueprint for Networking the Flow of Material, Information, and Cash (Resource Management)*. Boca Raton, FL: CRC Press.

Wall Street Journal article on clarity missing in the supply chain: http://online.wsj.com/article_email/SB124260855682928885-lMyQjAxMDI5NDEyODYxMDg4Wj.html.

REFERENCES

APICS Dictionary, http://www.apics.org/industry-content-research/publications/apics-dictionary

APICS, Operations Management Body of Knowledge (OMBOK) Framework, http://www.apics.org/industry-content-research/publications/ombok/apics-ombok-framework-table-of-contents

Carbon Disclosure Project (CDP) Supply Chain Report 2012. A New Era: Supplier Management in the Low Carbon Economy, https://www.cdproject.net/CDPResults/CDP-Supply-Chain-Report-2012.pdf

Department of Health and Services (USA), Office of the Inspector General, Traceability in the Food Supply Chain, http://oig.hhs.gov/oei/reports/oei-02-06-00210.pdf

FDA, Generally Regarded as Safe (GRAS), http://www.fda.gov/Food/FoodIngredients Packaging/GenerallyRecognizedasSafeGRAS/default.htm

FDA, Nanotechnology Taskforce Report 2007, http://www.fda.gov/ScienceResearch/Special Topics/Nanotechnology/UCM2006659.htm

FDA, Pathway to Global Product Safety and Quality, http://www.fda.gov/downloads/AboutFDA/CentersOffices/OC/GlobalProductPathway/UCM259845.pdf

GHG Protocol Scope 3, http://pdf.wri.org/ghgp_corporate_value_chain_scope_3_standard.pdf

Handy, C., *The Age of Unreason,* Harvard Business Review Press, 1989, pp. 137, 291 (quiz answers) and 304 (recommended reading).

Hoekstra, Arjen Y., Chapagain, Askok K., Aldaya, Maite M. and Mekonnen, Meskin M., *The Water Footprint Assessment Manual: Setting the Global Standard*, Routledge; Taylor & Francis Group, Oxford, UK, 2011, p. 206.

McLeod, S.A.; *Maslow's Hierarchy of Needs,* 2007, retrieved from http://www.simplypsychology.org/maslow.html

Moore, G. A., *Crossing the Chasm: Marketing and Selling High-Tech Products to Mainstream Customers,* Harper Business Essentials, New York, NY, 2002, pp. 190, 305.

Pine, J. B. II, *Mass Customization: The New Frontier in Business Competition,* Harvard Business School Press, Boston, MA, 1993, pp. 26, 277.

Registrar Corp., Number of registered food facilities registered in the USA, http://www.registrarcorp.com/fda-food/registration/index.jsp?lang=en

Supply-Chain Operations Reference model (SCOR), http://supply-chain.org/resources/scor

United Nations Global Compact, Supply Chain Sustainability: A Practical Guide for Continuous Improvement, http://www.unglobalcompact.org/docs/issues_doc/supply_chain/SupplyChainRep_spread.pdf

Wall Street Journal, Clarity Missing in Supply Chain, http://online.wsj.com/article/SB124260855682928885.html

Index

D

E

F

M

N

O

P

Q

S

T

U